Solving Math Quadratic Equations and Inequalities

by

Karen Kusanovich, Wendy Lawson, and Nghi H. Nguyen

DORRANCE PUBLISHING CO., INC.
PITTSBURGH PENNSYLVANIA 15222

All Rights Reserved
Copyright © 2004 by Karen Kusanovich, Wendy Lawson, and Nghi H. Nguyen
No part of this book may be reproduced or transmitted
in any form or by any means, electronic or mechanical,
including photocopying, recording, or by any information
storage and retrieval system without permission in
writing from the publisher.

ISBN # 0-8059-6682-X
Printed in the United States of America

First Printing

For information or to order additional books, please write:
Dorrance Publishing Co., Inc.
701 Smithfield Street
Third Floor
Pittsburgh, Pennsylvania 15222
U.S.A.
1-800-788-7654

Or visit our website and online catalogue at *www.dorrancepublishing.com*

PREFACE

This booklet is intended to be a supplement to traditional math courses. It covers a variety of methods for solving equations and inequalities that are core subjects found in traditional Algebra I and Algebra II classes.

It introduces a new method for solving quadratic equations quickly, recently developed by the authors. This method is a convenient approach which can replace the existing trial-and-error method for factoring that is taught in most Algebra I classes.

Innovative number line techniques and graphing approaches are also introduced to simplify solving quadratic inequalities and systems of quadratic inequalities. With the help of graphing calculators, students can use the graphical approach to solve complex problems with more conveniences than using traditional algebraic manipulations.

The wide variety of problems covered by this booklet make it an excellent reference for high school students who are looking for alternative methods and a deeper understanding of the techniques they have already learned.

THE AUTHORS

Karen Kusanovich	BA. Math., MA. in Education Leadership, San Jose State University, Ca. Math Department Chair and Math teacher
Wendy Lawson	BS. Math., MA. in Education, Stanford University, Ca. Math teacher.
Nghi H Nguyen	Graduated in Civil Engineering National University of Technology, Saigon City, Vietnam Math tutor.

SOLVING QUADRATIC EQUATIONS & INEQUALITIES IN ONE VARIABLE
Booklet Content

	Page
Chapter 1. Solving quadratic equations in one variable	
1-1. Polynomial equations	1
1-2. Generalities about solving quadratic equations in one variable.	
1-3. Solving by the quadratic formulas	
1-3-1. General formulas	2
1-3-2. Simplified formulas when b is an even number	4
1-4. Solving by factoring	6
1-5. Quick solving of quadratic equations	
1-5-1. Some tips for solving a quadratic equation quickly	7
1-5-2. The rule of sign for real roots	8
1-6. Solving a quadratic equation by the Diagonal Sum Method	9
1-6-1. Development of the Diagonal Sum Method	10
1-6-2. Practical operation steps of the Diagonal Sum Method	11
1-7. Some applications of the Diagonal Sum Method	
1-7-1. Application 1: Find a quadratic equation knowing the two roots	16
1-7-2. Application 2: Solving problems	21
1-7-3. Application 3: Checking answers of a quadratic equation	24
1-8. Chapter Review and Review Exercises	26
Chapter 2. Solving quadratic equations – Special cases	
2-1. Solving the quadratic equation $x^2 + bx + c = 0$	29
2-1-1. Solving by the quadratic formulas	30
2-1-2. Quick solving by mental math	31
2-1-3. Solving by the Diagonal Sum method	32
2-2. Solving by completing the square	34
2-3. Quadratic equations with parameters	35
2-4. Solving biquadratic equations	38

	Page
2-5. Equations with rational expressions.	39
2-6. Equations with radical expressions.	40
2-7. Chapter Review	43
2-8. Review exercises	45

Chapter 3. Solving quadratic inequalities and systems of quadratic inequalities

3-1. Generalities about solving quadratic inequalities

    3-1-1. Standard form of a quadratic inequality   48
    3-1-2. Solving approaches

3-2. Solving quadratic inequalities by the number line   49
3-3. Solving quadratic inequalities by the rule of sign

    3-3-1. Relationship between signs and roots of a trinomial   50
    3-3-2. Theorem on the sign status of a trinomial   51
    3-3-3. The rule of sign and the sign table of a trinomial
    3-3-4. Application of the rule of sign and the sign table   52

3-4. Solving a system of quadratic inequalities

    3-4-1. Solving by using the number line   53
    3-4-2. Solving by using the sign table.   55

3-5. Solving other systems of inequalities   58

    3-5-1. Solving a mixed system of linear and quadratic inequalities   59
    3-5-2. Solving a system of three or more quadratic inequalities
    3-5-3 Solving a system that can be transformed   60
           into quadratic inequalities

3-6. Chapter Review   63
3-7. Review exercises   64

Chapter 4. Solving quadratic inequalities in one variable by graphing

4-1. The quadratic function $y = f(x) = ax^2 + bx + c$ 67
4-2. Solving a quadratic inequality by graphing   72
4-3. Solving a system of quadratic inequalities by graphing   74
4-4. Solving the quadratic inequality $y \gtrless ax^2 + bx + c$ 79
4-5. Solving the system of quadratic inequalities $[y \gtrless f(x)\,; y \gtrless g(x)]$ 81
4-6. Review exercises   83

CHAPTER 1
SOLVING QUADRATIC EQUATIONS IN ONE VARIABLE

1-1. POLYNOMIAL EQUATIONS.

A polynomial equation is an equation in which both sides are polynomials of variable x. Polynomial equations are usually named by the term of highest degree of the variable's exponent. If a is a number different than zero and b, c, d are constants, then

$ax + b = 0$ is a linear equation

$ax^2 + bx + c = 0$ is a quadratic equation or second degree equation

$ax^3 + bx^2 + cx + d = 0$ is a cubic equation or third degree equation

1-2. GENERALITIES ABOUT SOLVING QUADRATIC EQUATION.

Solving a quadratic equation means finding a solution set or values of the variable x that make the equation true. Usually, the first step of solving is to transform the equation into standard form in which one side is zero. The other side is a simplified polynomial arranged in the order of decreasing degree of the variable's exponent. The standard form of a quadratic equation is $ax^2 + bx + c = 0$ where $a \neq 0$
The equation has a solution set with at maximum two real number roots or two values of x that make the equation true. Depending on the values of constants a, b, c, there may be one double root or no real root at all.

Example 1. The quadratic equation $x^2 - 3x + 2 = 0$ has the solution set (1, 2) which are the two real roots making the equation true.

Example 2. The equation $x^2 + 4x + 4 = 0$ has (-2) as a unique root, called a double root, which is the real value of x making the equation true.

Example 3. The equation $3x^2 + x + 5 = 0$ does not have any real roots.

Finding the solution set, or roots, is called solving the equation. There are many ways to solve quadratic equations. The general method uses formulas, called The Quadratic Formulas. Other approaches use mental math combined with trial and error to find the answers quickly.

1-3. SOLVING BY THE QUADRATIC FORMULAS

1-3-1. The quadratic formulas

The standard form of a quadratic equation is $ax^2 + bx + c = 0$ in which a, b and c are real numbers and $a \neq 0$

$$x^2 + \frac{b}{a}x + \frac{c}{a} = 0 \qquad \text{Divide both sides by } a$$

$$x^2 + \frac{b}{a}x = -\frac{c}{a} \qquad \text{Moving term}$$

$$x^2 + \frac{b}{a}x + \frac{b^2}{4a^2} = \frac{b^2}{4a^2} - \frac{c}{a} \qquad \text{Add } \frac{b^2}{4a^2} \text{ to both sides}$$

$$(x + \frac{b}{2a})^2 = \frac{b^2 - 4ac}{4a^2} \qquad \text{Binomial identity } (m+n)^2$$

$$x + \frac{b}{2a} = \pm \frac{\sqrt{D}}{2a}$$

with D, called the **Discriminant**, $D = b^2 - 4ac$

    If $D > 0$ There are two different real roots

$$x_1 = \frac{-b + \sqrt{D}}{2a} \qquad \text{and} \qquad x_2 = \frac{-b - \sqrt{D}}{2a}$$

    If $D = 0$ There is one double real root, $x_1 = x_2 = -\frac{b}{2a}$

    If $D < 0$ There are no real roots.
        Note: However, there are two imaginary roots in the form of two conjugate complex numbers that we will learn later.

Sum and product of roots

Sum $S = x_1 + x_2 = -\dfrac{b}{a}$

Product: $P = x_1 \cdot x_2 = \dfrac{c}{a}$

<u>Example 1</u>. Solve the equation $2x^2 + 7x - 4 = 0$

Solution. Discriminant $D = b^2 - 4ac = (7)^2 - 4(2)(-4) = 49 + 32 = 81 = 9^2$

The equation has two real roots

$$x = \dfrac{-7 + 9}{4} = \dfrac{1}{2} \qquad x = \dfrac{-7 - 9}{4} = \dfrac{-16}{4} = -4$$

<u>Example 2</u>. Solve the equation $5x^2 + 9x - 2 = 0$

Solution. Discriminant $D = 81 + 40 = 121 = 11^2$

The equation has two real roots

$$x = \dfrac{-9 + 11}{10} = \dfrac{1}{5} \qquad x = \dfrac{-9 - 11}{10} = -2$$

<u>Example 3</u>.   Solve the equation $4x^2 - 12x + 9 = 0$

Solution. Discriminant $D = (-12)^2 - 4(9)(4) = 0$

The equation has a double root $x_1 = x_2 = \dfrac{12}{8} = \dfrac{3}{2}$

<u>Example 4</u>   Solve the equation $2x^2 - 3x + 7 = 0$

Solution. $D = 9 - 56 = -47$ The equation does not have real roots.

Exercises on solving by quadratic formulas

1. $5x^2 + 9x - 2 = 0$ 2. $6x^2 + 13x + 5 = 0$

3. $2x^2 + 7x - 4 = 0$ 4. $10x^2 + 11x - 35 = 0$

5. $6x^2 + 15x + 9 = 0$ 6. $6x^2 + x - 35 = 0$

7. $21x^2 - 5x - 6 = 0$ 8. $4x^2 + 5x + 1 = 0$

9. $3x^2 - 5x + 1 = 0$ 10. $2x^2 + 3x - 1 = 0$

11. $3x^2 + 5x + 1 = 0$ 12. $x^2 + 5x + 5 = 0$

13. $7x^2 - 23x + 6 = 0$ 14. $5x^2 - 13x + 6 = 0$

1-3-2. Simplified quadratic formulas when b is an even number

When b is an even number we have simplified quadratic formulas. We encourage students to learn and use these simplified formulas because they apply to 50% of the cases.

When b is an even number, $b = 2b'$ or $b' = \dfrac{b}{2}$

The Discriminant $D = b^2 - 4ac = (2b')^2 - 4ac = 4(b'^2 - ac)$

Call **D'** the new Discriminant   **$D' = b'^2 - ac$**.   with $D = 4D'$

The quadratic formulas become simplified:

If $D' > 0$ There are two real roots
$$x_1 = \dfrac{-b' + \sqrt{D'}}{a} \quad \text{and} \quad x_2 = \dfrac{-b' - \sqrt{D'}}{a}$$

If $D' = 0$ There is one double root $x_1 = x_2 = -\dfrac{b'}{a}$

If $D' < 0$ There are no real roots.

<u>Example 1.</u>   Solve the equation $3x^2 + 16x - 12 = 0$

Solution. Discriminant $D' = (8)^2 + 3 \cdot 12 = 64 + 36 = 100 = 10^2$

Two real roots $\dfrac{-8 \pm 10}{3}$. They are $(\dfrac{2}{3})$ and (-6)

Example 2.   Solve the equation $7x^2 + 18x - 25 = 0$

Solution. Discriminant $D' = 9^2 - (7)(-25) = 81 + 175 = 256 = 16^2$

Two real roots $\dfrac{-9 \pm 16}{7}$. They are (1) and $(-\dfrac{25}{7})$

Example 3.   Solve the equation $4x^2 - 12x + 9 = 0$

Solution. Discriminant $D' = 36 - 36 = 0$

There is one double root $x_1 = x_2 = \dfrac{6}{4} = \dfrac{3}{2}$

Example 4.   Solve the equation $2x^2 + 12x + 17 = 0$

Solution. Discriminant $D' = 36 - 34 = 2$

Two real roots $\dfrac{-6 \pm \sqrt{2}}{2}$. They are $(-3 + \dfrac{\sqrt{2}}{2})$ and $(-3 - \dfrac{\sqrt{2}}{2})$

Exercises on solving by quadratic formulas (b even number)

1. $3x^2 + 4x - 4 = 0$
2. $8x^2 + 22x - 21 = 0$
3. $16x^2 - 34x - 15 = 0$
4. $19x^2 - 18x - 1 = 0$
5. $3x^2 - 8x - 11 = 0$
6. $15x^2 - 4x - 35 = 0$
7. $4x^2 + 4x - 5 = 0$
8. $8x^2 - 22x + 15 = 0$
9. $3x^2 - 16x + 21 = 0$
10. $3x^2 - 6x - 1 = 0$
11. $5x^2 - 10x - 3 = 0$
12. $7x^2 + 8x + 2 = 0$

1-4. SOLVING QUADRATIC EQUATIONS BY FACTORING.

The factoring method uses mental math combined with trial and error to factor the left side of a quadratic equation that is in standard form:

$$ax^2 + bx + c = 0 \qquad (a \neq 0) \qquad (1)$$

Suppose a_1, a_2 and c_1, c_2 are a factor set of constants a and c

$$a = a_1 \cdot a_2 \qquad \text{and} \qquad c = c_1 \cdot c_2$$

The quadratic equation can be written in the factored form:

$$(a_1 x - c_1)(a_2 x - c_2) = 0 \qquad (2)$$

If these factors are the correct ones, then the distributive multiplication of equation (2) will give exactly the terms of equation (1). If the terms do not match, students are to try other possible factor sets of a and c until the correct factors are found.
This booklet does not develop in detail this factoring method since students can find it in other Algebra books.

Note. Through experience and with a sharp mental math ability, students can manage to get the answer faster through factoring than by using the quadratic formulas. However, the ultimate goal is to quickly find the roots. So if the factoring operation gets confusing or takes too much time, students would be better off going back to the quadratic formulas.
Another suggestion is to mentally estimate the Discriminant (D or D') to see if it is positive before solving.

<u>Example 1.</u> Solve by factoring $\qquad 8x^2 - 22x + 15 = 0$

Solution. The selected factor sets for a are (4, 2), (2, 4) and for c are (3, 5), (5, 3), (-3, -5). The probable factors of the equation are written below. After a few trial and error computations, the answer can be found.

$(2x + 3)(4x + 5) = 0$	$8x^2 + 10x + 12x + 15$	No
$(2x + 5)(4x + 3) = 0$	$8x^2 + 6x + 20x + 15$	No
$(4x - 3)(2x - 5) = 0$	$8x^2 - 20x - 6x + 15$	No
$(4x - 5)(2x - 3) = 0$	$8x^2 - 12x - 10x + 15$	Yes

After the factored form is found, solve the two binomials for x

$$4x - 5 = 0 \qquad\qquad 2x - 3 = 0$$
$$x = \frac{5}{4} \qquad\qquad x = \frac{3}{2}$$

<u>Example 2</u>. Solve by factoring $6x^2 - x - 15 = 0$

Solution. Possible factor sets for a $(2, 3), (3, 2)$
　　　　Possible factor set for c $(5, -3), (3, -5)$

Possible factors for the equation:

$(3x + 3)(2x - 5) = 0$	$6x^2 - 15x + 6x - 15$	No
$(3x - 3)(2x + 5) = 0$	$6x^2 + 15x - 6x - 15$	No
$(2x - 3)(3x + 5) = 0$	$6x^2 + 15x - 9x - 15$	No
$(2x + 3)(3x - 5) = 0$	$6x^2 - 10x + 9x - 15$	Yes

The correct factoring is $(2x + 3)(3x - 5) = 0$. Solve the binomials for x

$$2x + 3 = 0 \qquad\qquad 3x - 5 = 0$$
$$x = -\frac{3}{2} \qquad\qquad x = \frac{5}{3}$$

1-5. QUICK SOLVING OF QUADRATIC EQUATIONS.

Solving by quadratic formulas is necessary when the Discriminant D is not a perfect square or when the square root of D is an irrational number. In other cases, we can solve a quadratic equation faster than using the quadratic formulas. We introduce here a new method, recently developed by the authors, using mental math to quickly find the real roots. Before learning this method, it is necessary that students know about the rule of sign for real roots and some tips for solving a quadratic equation quickly.

1-5-1. Some Tips for solving a quadratic equation quickly.

There are a few tips that will help students in solving quadratic equations quickly.

Tip 1. If a is a negative number, it is helpful to factor out (-1) as a common factor to make computation easier and to avoid mistakes.

Example.
$$-7x^2 + 9x - 2 = 0$$
$$(-1)(7x^2 - 9x + 2) = 0$$

Tip 2. If the sum of constants $a + b + c = 0$ then one of the roots is (1) and the other root is ($\frac{c}{a}$)

It is true because $\quad a(1) + b(1) + c = 0$

Example 1. The equation $\quad 3x^2 - 5x + 2 = 0$
has one root $x = 1$ and the other is $x = \frac{2}{3}$

Tip 3. If $a - b + c = 0$, then one root is (-1) and the other root is ($-\frac{c}{a}$).

Example 2 The equation $\quad x^2 + 7x + 6 = 0$
has two roots (-1) and (-6)

1-5-2. Rule of Sign for real roots.

To solve a quadratic equation quickly, it is helpful for students to know and remember a rule about the signs of the real roots. The rule is stated as follow:

1. If a and c are opposite signs (ac < 0), the product P of the two roots is negative. Consequently, the equation always has two real roots of opposite signs.

Example 3. Solve the equation $\quad 4x^2 + 24x - 13 = 0$

Solution. Constants a and c are opposite signs, so there are two real roots of opposite signs.

Discriminant $D' = (12)^2 - (4)(-13) = 144 + 52 = 196 = 14^2$

Two real roots $x_1 = \frac{1}{2}$ and $x_2 = -6\frac{1}{2}$

2. If a and c are the same sign (ac > 0), the product of the two roots is positive and consequently, the two real roots have the same sign.

    a. If a and b are the same sign, then the sum $(-\frac{b}{a})$ is negative, and the two real roots are both negative

    b If a and b are opposite signs, then the sum $(-\frac{b}{a})$ is positive, and the two real roots are both positive.

Example 4. Solve the equation $15x^2 + 22x + 8 = 0$

Solution. Since a and c are both positive, and a and b are the same sign, both roots are negative. After solving, they are $(-\frac{2}{3})$ and $(-\frac{4}{5})$

Example 5. Solve the equation $3x^2 - 13x + 10 = 0$

Solution. Since a and c are the same sign, the two real roots have the same sign. Since a and b are opposite signs, both real roots are positive: After solving, they are (1) and $(\frac{10}{3})$.

Example 6. The equation $x^2 + 7x + 6 = 0$ has two real roots -1 and -6, both negative

Example 7. The equation $x^2 + 13x + 30 = 0$ has two real roots -3 and -10, both negative

1-6. SOLVING A QUADRATIC EQUATION BY THE DIAGONAL SUM METHOD

    We particularly recommend students use the following method that combines mental math and the trial and error approach. This new method has the advantage of **giving the real roots** of the equation **directly** without having to factor it. This method does not require students to have a sharp mental math ability. Students just need to know how to list all possible root sets by permutation. Then they apply a simple formula to find out which set is the solution set.

We recommend students study this method thoroughly to fully master the process of solving and factoring a quadratic equation. Again, it is necessary that students master the Rule of Sign. See Paragraph 1-5-2.

1-6-1. Development of the Diagonal Sum Method.

The concept of the method is to find two fractions whose sum is $(-\frac{b}{a})$ and whose product is $(\frac{c}{a})$.

Equation to solve in standard form $\quad ax^2 + bx + c = 0 \quad (a \neq 0)$

Suppose (a_1) and (a_2) are any selected factors of coefficient a, $a = a_1 \cdot a_2$

and suppose (c_1) and (c_2) are any selected factors of constant c, $c = c_1 \cdot c_2$

The set of possible roots is in the form of two fractions $(\frac{c_1}{a_1}, \frac{c_2}{a_2})$

Their product is $\quad P = (\frac{c_1}{a_1})(\frac{c_2}{a_2}) = \frac{c}{a}$

By permutation, there are two sets of possible roots that are in the form of fractions

$$(\frac{c_1}{a_1}, \frac{c_2}{a_2}) \qquad (\frac{c_1}{a_2}, \frac{c_2}{a_1})$$

If the sum of any set of these two root sets is equal to the quotient $(-\frac{b}{a})$, then this set is the solution set.

Sum of roots $\quad \frac{c_1}{a_1} + \frac{c_2}{a_2} = \frac{c_1 a_2 + c_2 a_1}{a_1 a_2} = \frac{c_1 a_2 + c_2 a_1}{a} = -\frac{b}{a}$

The expression $(c_1 a_2 + c_2 a_1)$ is called the **diagonal sum** of any set of two fraction roots.

$$(\frac{c_1}{a_1} \times \frac{c_2}{a_2}) \qquad \text{Diagonal sum} \quad c_1 a_2 + c_2 a_1$$

Rule: Find the diagonal sum which is equal to $-b$ from all the possible root sets.

From this analysis, we can set up a practical approach to solve a quadratic equation when (a) and (c) are factorable numbers. If a and c are prime numbers, the factors are one and the number itself.

1-6-2. Practical operation steps of The Diagonal Sum Method.

a. First, look for shortcuts as guided by the **Tips** (See Paragraph 1-5-1). Then, use the **rule of sign** to know in advance the sign of the two roots. (See Paragraph 1-5-2)

b. Select one possible root set which is in the form of two fractions $(\frac{c_1}{a_1}, \frac{c_2}{a_2})$

c. By permutation, there are only two possible root sets if the two roots are same sign. Knowing the rule of sign helps limit the number of permutations. There are 4 possible root sets if a and c are opposite signs. However, we only have to write two of them and we will explain why.

d. Mentally compute the diagonal sum of each of the two fraction sets and compare each sum to **(-b)**. If it matches, the two related fractions are the real roots. Through experience and practice, students can quickly find the diagonal sum that fits.

e. If the diagonal sum is neither **(b)** or **(-b)**, select other factors of c and a and repeat the steps again. In fact, a brief mental study of the sum $(-\frac{b}{a})$ may show which combination of $(\frac{c}{a})$ to select.

<u>Example 1.</u>   Solve the equation $6x^2 + x - 12 = 0$

Solution. Since a and c are opposite signs, the roots are opposite signs.

Select (3, 4) as factors of 12 and select (2, 3) as factors of 6.

By permutation, list the set of possible fraction roots. There are 4 of them.

$(\frac{-3}{2}, \frac{4}{3})$ $(\frac{-3}{3}, \frac{4}{2})$ $(\frac{3}{2}, \frac{-4}{3})$ $(\frac{3}{3}, \frac{-4}{2})$

Mentally compute diagonal sum $c_1a_2 + c_2a_1$

8 - 9 = -1 = **- b**   12 – 6 = 6   9 – 8 = 1   6 – 12 = -6

The real roots are fractions $(-\frac{3}{2})$ and $(\frac{4}{3})$

11

Important Remark: The four diagonal sums are **opposite**, two by two. So, practically, we only have to write and to mentally compute the diagonal sum of the **first two** fraction sets. The other two sums can be obtained by taking the opposite of the first two sets, if necessary.

Example 2.   Solve the equation $6x^2 - 11x - 35 = 0$

Solution. The two roots are opposite signs since a and c are opposite signs.

Select $\frac{c}{a}$ permutation sets. Write only **two** of them

$(\frac{5}{2}, \frac{-7}{3})$ $(\frac{5}{3}, \frac{-7}{2})$

Mentally compute the diagonal sum of the 2 sets.

(1)   (-11) = b

The real roots are opposite to the fractions in the second set. They are the fractions $(-\frac{5}{3})$ and $(\frac{7}{2})$

Check by factoring if necessary $(3x + 5)(2x - 7) = 0$

Example 3.   Solve the equation $6x^2 + 29x + 35 = 0$

Solution. Both roots are negative since all constants are positive.

Note: The rule of sign helps limit the possible root sets to two.

$(\frac{-5}{2}, \frac{-7}{3})$ $(\frac{-5}{3}, \frac{-7}{2})$

Mentally compute the diagonal sum:

(-29) = - b   (-31)

The real roots are fractions $(-\frac{5}{2})$ and $(-\frac{7}{3})$

Example 4.   Solve the equation $-6x^2 + 7x + 20 = 0$ (1)

Solution. Factor out (-1) $(-1)(6x^2 - 7x - 20) = 0$ (2)

Select c/a sets (roots are opposite signs):

$(\frac{-4}{2}, \frac{5}{3})$ $(\frac{-4}{3}, \frac{5}{2})$

Mentally compute the diagonal sum.
Stop doing it once you get either (b) or (-b)

   (-2)   (7) = - b

Real roots are fractions $(-\frac{4}{3})$ and $(\frac{5}{2})$

Note. The two fractions are real roots of **both equations** (1) and (2).

Check by factoring if being asked: $(-1)(3x + 4)(2x - 5) = 0$

Example 5. Solve $-4x^2 + 39x - 56 = 0$

Solution. Factor out $(-1)(4x^2 - 39x + 56) = 0$

Since a and c are the same sign, both roots are the same sign. Since a and b are different signs, both roots are positive.

Select c/a permutation sets

$(\frac{7}{1}, \frac{8}{4})$ $(\frac{7}{4}, \frac{8}{1})$

Diagonal sum

   (36)   **(39) = - b**   (30)

The two roots are $(\frac{7}{4})$ and (8).

Note. This method is particularly convenient when the constants a, b, c are big numbers and/or when a, c are **prime** numbers.

Example 6. Solve $-6x^2 + 19x + 77 = 0$

Solution. Factor out $(-1)(6x^2 - 19x - 77) = 0$

Two roots are opposite signs. Select (c/a) permutation sets

$(\frac{-7}{2}, \frac{11}{3})$ $(\frac{-7}{3}, \frac{11}{2})$

The real roots are the fractions $(-\frac{7}{3})$ and $(\frac{11}{2})$.

Example 7. Solve $7x^2 - 118x - 17 = 0$

The roots are opposite sign. Both numbers 7 and -17 are prime.

$(\frac{1}{7}, \frac{-17}{1})$ (See Note).

$-119 + 1 = -118 = b$

The real roots have opposite signs to the fractions of the set. They are $-\frac{1}{7}$ and 17

Note. The second set can be ignored because this is not the case of **Tip 1**. The equation does not have 1 as real root.

Example 8. Solve $21x^2 + 50x + 24 = 0$

Solution. Both roots are negative. First, try the two sets $(\frac{-4}{3}, \frac{-6}{7})$ $(\frac{-4}{7}, \frac{-6}{3})$

Neither diagonal sum is -50.

Next, try the two sets $(\frac{-2}{3}, \frac{-12}{7})$ $(\frac{-2}{7}, \frac{-12}{3})$

Diagonal sum $-14 - 36 = -50$

The real roots are the two fractions of the first set.

REMARK. When $a = 1$, the quadratic equation is reduced to $x^2 + bx + c = 0$. The Diagonal Sum Method reduces itself to a popular puzzle: finding two number whose sum is $(-b)$ and whose product is (c). The puzzle solving will be developed in the next chapter.

Exercises on solving quadratic equations by the Diagonal Sum Method.

Solve these equations by using the Diagonal Sum method.

1. $2x^2 + 7x - 4 = 0$
2. $5x^2 + 9x - 2 = 0$
3. $6x^2 + 13x + 5 = 0$
4. $6x^2 - x - 15 = 0$
5. $8x^2 + 22x - 21 = 0$
6. $10x^2 + 11x - 35 = 0$
7. $35x^2 - 11x - 6 = 0$
8. $9x^2 - 11x - 20 = 0$
9. $6x^2 + 15x + 9 = 0$
10. $15x^2 + 22x + 8 = 0$
11. $12x^2 - 17x - 7 = 0$
12. $12x^2 - 59x - 5 = 0$
13. $8x^2 + 35x + 38 = 0$
14. $21x^2 - 23x + 6 = 0$
15. $16x^2 - 34x - 15 = 0$
16. $6x^2 - 11x - 35 = 0$
17. $21x^2 - 5x - 6 = 0$
18. $-6x^2 + 19x + 77 = 0$
19. $6x^2 - x - 35 = 0$
20. $21x^2 - 5x - 6 = 0$

More exercises on solving by the Diagonal Sum Method.

1. $35x^2 + 61x + 24 = 0$
2. $35x^2 + 179x + 20 = 0$
3. $20x^2 - 59x + 42 = 0$
4. $-21x^2 + 13x + 20 = 0$
5. $15x^2 + x - 40 = 0$
6. $8x^2 - 22x + 15 = 0$
7. $21x^2 - 23x + 6 = 0$
8. $16x^2 - 34x - 15 = 0$
9. $6x^2 + x - 35 = 0$
10. $-4x^2 + 39x - 56 = 0$
11. $3x^2 + 2x - 8 = 0$
12. $15x^2 + 94x - 40 = 0$
13. $11x^2 - 120x - 11 = 0$
14. $17x^2 + 324x + 19 = 0$
15. $13x - 222x + 17 = 0$
16. $7x^2 - 92x + 13 = 0$

1-7. SOME APPLICATIONS OF THE DIAGONAL SUM METHOD

1-7-1. Application 1.

Given two fractions $(\frac{c_1}{a_1})$ and $(\frac{c_2}{a_2})$. They are the two roots of a quadratic equation in standard form $ax^2 + bx + c = 0$ in which
$c = c_1 c_2$, $a = a_1 a_2$, and $-b = (c_1 a_2 + c_2 a_1)$

a. <u>When the roots are real number fractions.</u> In this case, D (or D') is a perfect square number and $\sqrt{D}$ (or $\sqrt{D'}$) is a rational number.

<u>Example 1</u>. Find a quadratic equation in standard form $ax^2 + bx + c = 0$ knowing that the two real roots are two fractions $(\frac{3}{2})$ and $(\frac{7}{4})$

Solution:

$a = (2)(4) = 8$ $a = a_1 a_2$
$c = (3)(7) = 21$ $c = c_1 c_2$
$-b = (3 \cdot 4 + 7 \cdot 2) = 26$ Diagonal sum $a_1 c_2 + a_2 c_1$

A quadratic equation in standard form is

$8x^2 - 26x + 21 = 0$ Answer

<u>Example 2</u>. Find a quadratic equation in standard form which has two fractions $(\frac{5}{8})$ and $(-\frac{7}{9})$ as real roots.

Solution:

$a = (8)(9) = 72$ $a = a_1 a_2$
$c = ((5)(-7) = -35$ $c = c_1 c_2$
$-b = (5 \cdot 9 - 7 \cdot 8) = -11$ Diagonal sum $a_1 c_2 + a_2 c_1$

An equation is

$72x^2 + 11x - 35 = 0$ Answer

Check: Using The Diagonal Sum Method, the real roots are $(\frac{5}{8})$ and $(-\frac{7}{9})$

Example 3. Find a quadratic equation in standard form which has $(-\frac{5}{4})$ and $(-\frac{7}{6})$ as real roots

Solution.
$$c = (-5)(-7) = 35 \qquad c = c_1 c_2$$
$$a = (4)(6) = 24 \qquad a = a_1 a_2$$
$$-b = (-30 - 28) = -58 \qquad \text{Diagonal sum}$$

A quadratic equation is

$$24x^2 + 58x + 35 = 0 \qquad \text{Answer}$$

b. When the roots are fractions with radical expressions. When the square root of the Discriminant D (or D') is an irrational number, the two roots are two fractions which contain a radical expression. Application 1 also works in this case.

Example 4. Given two fractions with a radical $\frac{1+\sqrt{3}}{2}$ and $\frac{1-\sqrt{3}}{2}$.

Find a quadratic equation in standard form that has these fractions as two real roots.

Solution.
$$a = (2)(2) = 4 \qquad a = a_1 a_2$$
$$c = (1+\sqrt{3})(1-\sqrt{3}) = 1 - 3 = -2 \qquad c = c_1 c_2 \text{ . Identity } (a+b)(a-b)$$
$$-b = (2 + 2\sqrt{3}) + (2 - 2\sqrt{3}) = 4 \qquad \text{Diagonal sum } a_1 c_2 + a_2 c_1$$

An equation is $\quad 4x^2 - 4x - 2 = 0$

or $\quad 2x^2 - 2x - 1 = 0` \qquad$ Answer

Check by solving the equation, using the quadratic formulas:

D' = 1 + 2 = 3. Two real roots $\frac{1+\sqrt{3}}{2}$ and $\frac{1-\sqrt{3}}{2}$

Example 5. Find a quadratic equation in standard form which has $\dfrac{2+\sqrt{3}}{2}$ and $\dfrac{2-\sqrt{3}}{2}$ as two real roots

Solution

$a = (3)(3) = 9$ $a = a_1 a_2$

$c = (2+\sqrt{3})(2-\sqrt{3}) = 4 - 3 = 1$ $c = c_1 c_2$ Identity $(a+b)(a-b)$

$-b = (6 + 3\sqrt{3}) + (6 - 3\sqrt{3}) = 12$ Diagonal sum

An equation is

$9x^2 - 12x + 1 = 0$ Answer

Check by solving the equation, using the quadratic formulas:

$D' = 36 - 9 = 27 = 9 \cdot 3$

Two roots $\dfrac{2+\sqrt{3}}{3}$ and $\dfrac{2-\sqrt{3}}{3}$.

     c. **When the roots are fractions with complex numbers.** When the <u>Discriminant</u> D (or D') is negative, there are no real roots. However there are two complex roots which are two fractions in the form of two complex numbers.
     A complex number is a number of the form $a + bi$, where a and b stand for real number and i stands for an imaginary number unit with $i^2 = -1$ (or $i = \pm\sqrt{-1}$)
     In this case, D (or D') < 0, the two roots are in the form of two complex numbers that are conjugates of each other.

For example, the quadratic equation $3x^2 - 4x + 2 = 0$ has its discriminant $D' = 4 - 6 = -2 = 2i^2$

Its two roots are two conjugate complex numbers $\dfrac{2 \pm i\sqrt{2}}{3}$

Application 1 also works in this case. We will study more about imaginary numbers and complex numbers later.

Example 6. Find a quadratic equation in standard form which has as roots the two fractions in the form of two conjugate complex number $\dfrac{-2 \pm 5i}{3}$

$a = (3)(3) = 9$ $a = a_1 a_2$
$c = (-2 + 5i)(-2 - 5i) = 25 + 4 = 29$ $c = c_1 c_2$ Identity $(a + b)(a - b)$
$-b = (-6 + 5i) + (-6 - 5i) = -12$ Diagonal sum

An equation is

$$9x^2 + 12x + 29 = 0 \qquad \text{Answer}$$

Check by solving the equation, using the quadratic formula:

$D' = 36 - 261 = -225 = (15i)^2$ With $i^2 = -1$

Two roots $\dfrac{-6 \pm 15i}{9} = \dfrac{-2 \pm 5i}{3}$

Example 7. Two roots are conjugate complex numbers $\dfrac{3 \pm 11i}{10}$
Find a quadratic equation in standard form.

Solution.

$a = (10)(10) = 100$ $a = a_1 a_2$
$c = (3 + 11i)(3 - 11i) = 9 + 121 = 130$ $c = c_1 c_2$
$-b = (30 - 110i) + (30 + 110i) = 60$ Diagonal sum

An equation is
$$100x^2 - 60x + 130 = 0$$
$$10x^2 - 6x + 13 = 0 \qquad \text{Divide by 10. Answer}$$

Check: $D = 9 - 130 = -121 = (11i)^2$ With $i^2 = -1$

Two roots $\dfrac{3 + 11i}{10}$ and $\dfrac{3 - 11i}{10}$.

FINAL REMARK.

In general, a quadratic equation $ax^2 + bx + c = 0$ always has two roots depending on the values of the constants a, b, and c. The two roots are in the form of two fractions $(\frac{c_1}{a_1})$ $(\frac{c_2}{a_2})$ whose values are

$(\frac{-b \pm \sqrt{D}}{2a})$ or $(\frac{-b' \pm \sqrt{D'}}{a})$ where D (or D') is the Discriminant.

- If $\sqrt{D}$ (or $\sqrt{D'}$) is a rational number, then c_1 and c_2 are real numbers and the two roots are real number fractions. For example, the two roots are

$(\frac{2}{7})$ and $(\frac{3}{5})$

- If $\sqrt{D}$ (or $\sqrt{D'}$) is an irrational number, then c_1 and c_2 are numbers with radical expression. For example, the two roots are $\frac{2+\sqrt{3}}{3}$ and $\frac{2-\sqrt{3}}{3}$

- If D (or D') is a negative number, then c_1 and c_2 are conjugate complex numbers. For example, the two roots are $(\frac{3+2i}{4})$, $(\frac{3-2i}{4})$

Exercises on Application 1. Find a quadratic equation in standard form whose roots are

1. $-\frac{11}{7}$, $\frac{7}{9}$
2. $-\frac{5}{7}$, $\frac{9}{3}$
3. $-\frac{3}{8}$, $\frac{5}{2}$
4. $-\frac{9}{7}$, $\frac{3}{5}$
5. $-\frac{7}{4}$, 8
6. $\frac{9}{5}$, $\frac{3}{7}$
7. $-\frac{8}{7}$, $-\frac{9}{3}$
8. $-\frac{5}{3}$, $-\frac{1}{2}$
9. $\frac{2+\sqrt{5}}{3}$, $\frac{2-\sqrt{5}}{3}$
10. $\frac{3+\sqrt{7}}{4}$, $\frac{3-\sqrt{7}}{4}$
11. $\frac{2+\sqrt{3}}{5}$, $\frac{2-\sqrt{3}}{5}$
12. $\frac{4+\sqrt{13}}{3}$, $\frac{4-\sqrt{13}}{3}$
13. $\frac{3+\sqrt{5}}{4}$, $\frac{3-\sqrt{5}}{4}$
14. $\frac{5+\sqrt{6}}{7}$, $\frac{5-\sqrt{6}}{7}$
15. $\frac{2+3i}{5}$, $\frac{2-3i}{5}$
16. $\frac{2+5i}{4}$, $\frac{2-5i}{4}$
17. $\frac{3+4i}{2}$, $\frac{3-4i}{2}$
18. $\frac{3+7i}{4}$, $\frac{3-7i}{4}$
19. $\frac{3+5i}{4}$, $\frac{3-5i}{4}$
20. $\frac{5+3i}{7}$, $\frac{5-3i}{7}$

1-7-2. Application 2. Solving problems with quadratic equations.

<u>Problem 1.</u> Find an integer knowing that seven times the sum of itself and its reciprocal is 50.

Solution. Let x be the integer. Its reciprocal is $\frac{1}{x}$. The main equation is

$7(x + \frac{1}{x}) = 50$ This equation is undefined when $x = 0$

$7x^2 - 50x + 7 = 0$ Multiply both sides by x then simplify.

Two roots $(\frac{1}{7})$ and $(\frac{7}{1})$ Solving by Diagonal Sum method
                                           The integer is 7. Answer

Check: $7(7 + \frac{1}{7}) = 49 + 1 = 50$

Remark. Given a quadratic equation $ax^2 + bx + c = 0$. If $a = c$, $\frac{c}{a} = 1$, the two real roots are **reciprocal**. Their product is equal to one.

If $-b = a^2 + 1$, then one real root is a and the other is $\frac{1}{a}$.

Examples:

The equation $5x^2 - 26x + 5 = 0$ has two real roots   5 and $\frac{1}{5}$

The equation $11x^2 - 122x + 11 = 0$ has two real roots   11 and $\frac{1}{11}$

The equation $13x^2 - 170x + 13 = 0$ has two real roots   13 and $\frac{1}{13}$

<u>Problem 2.</u> A boat goes upstream from A to B then comes back downstream to A. Total round trip travel time is six hours. The river's current velocity is 2 km/hr. The distance AB is 35 km. Find the boat's own velocity.

Solution. Let x be the boat's own velocity. Its upstream velocity is $x - 2$ and its downstream velocity is $x + 2$. The condition is x must be greater than the current's velocity, $x > 2$.

The upstream travel time is $\frac{35}{x - 2}$ and the downstream travel time is $\frac{35}{x + 2}$

Set up the main equation of the problem:

$$\frac{35}{x-2} + \frac{35}{x+2} = 6 \qquad \text{Round trip time equal to 6}$$
Conditions: $x > 2$

$35(x - 2) + 35(x + 2) = 6(x^2 - 4)$ Multiply both sides by $(x - 2)(x + 2)$

or $6x^2 - 70x - 24 = 0$ Quadratic equation

The real roots are $(-\frac{1}{3})$ and (12) Diagonal Sum method

The answer (12) is justified since it is greater than 2.

<u>Problem 3</u>. Find the two sides of a rectangle knowing that its perimeter is $\frac{148}{15}$ m and its area is $\frac{91}{15}$ m.

Solution. The sum of the two sides is $\frac{74}{15}$ and their product is $\frac{91}{15}$. The two sides are the two real roots of the quadratic equation

$$x - \frac{74x}{15} + \frac{91}{15} = 0 \qquad \text{Property of quadratic equation}$$

or $15x^2 - 74x + 91 = 0$ Quadratic equation in standard form

Two real roots $(\frac{7}{3})$ and $(\frac{13}{5})$ m. Answers

<u>Problem 4</u>. Find the fraction knowing that the sum of itself and its reciprocal is $\frac{113}{56}$

Solution. Let x be the fraction. Its reciprocal is $\frac{1}{x}$. Set up the main equation

$$x + \frac{1}{x} = \frac{113}{56} \qquad \text{Sum of itself and its reciprocal}$$
Condition $x \neq 0$

$56x^2 - 113x + 56 = 0$ Multiply both sides by $56x$

Real roots are $(\frac{7}{8})$ $(\frac{8}{7})$ Solving by Diagonal Sum method
Both fractions are answers.

Exercises on solving problems.

Problem 1. An airplane flies from City A to City B 297 km from A. The wind speed is 20 km/hr. Total round trip travel time is 3 hr. Find the speed of the airplane without wind.

Problem 2. In a right triangle one leg is 3 m shorter than the other leg. The square of the hypotenuse is $\frac{1445}{9}$. Find the two legs.

Problem 3. Find two fractions whose sum is $\frac{151}{12}$ and whose product is $\frac{475}{12}$

Problem 4. Find two fractions whose sum is $\frac{183}{20}$ and whose product is $\frac{391}{20}$

Problem 5. Find the two sides of a rectangle knowing that its perimeter is $\frac{113}{6}$ cm and its area is $\frac{253}{12}$ square cm.

Problem 6. Find the sides of a rectangle knowing that its perimeter is $\frac{752}{25}$ meter and its area is $\frac{903}{25}$ square meter.

Problem 7. Find a fraction, knowing that two times the sum of itself and its reciprocal is equal to $\frac{116}{21}$

Problem 8. Find an integer knowing that nine times the sum of itself and its reciprocal is equal to 82

Problem 9. Find a integer knowing that five times the difference of itself and its reciprocal is $\frac{400}{9}$

Problem 10. Find a fraction knowing that two times the difference of itself and its reciprocal is $\frac{-336}{13}$

Problem 11. Find a fraction knowing that the sum of itself and three times its reciprocal is $\frac{148}{7}$

1-7-3. Application 3. Checking answers when solving a quadratic equation

If you solve a quadratic equation by the quadratic formulas, you can check the answers by mental math. Use these simple relations:

$$a_1 a_2 = a \ ; \quad c_1 c_2 = c$$
$$a_1 c_2 + a_2 c_1 = -b$$

a. When the answers are two fractions.

Example 1. Check the equation $16x^2 - 34x - 15 = 0$ and its roots $(-\frac{3}{8}), (\frac{5}{2})$

Solution. $8(2) = 16 = a$ $5(-3) = -15 = c$
Diagonal sum $40 - 6 = 34 = -b$ The answers are correct.

Example 2. Check the equation $21x^2 - 5x - 6 = 0$ and its roots $(\frac{2}{3}), (-\frac{3}{7})$

Solution. $3(7) = 21 = a$ $2(-3) = -6 = c$
                 $14 - 9 = 5 = -b$ The answers are correct

Example 3. Check the roots $(\frac{5}{2}), (-\frac{7}{5})$ of the equation $10x^2 + 11x - 35 = 0$

Solution. $2(5) = 10 = a$ $5(-7) = -35 = c$
                 $25 - 14 = 11 = b$ The answers are incorrect.
The correct ones are $(-\frac{5}{2})$ and $(\frac{7}{5})$

Example 4. Check the roots $(-17), (\frac{1}{7})$ of the equation $7x^2 - 118x - 17 = 0$

Solution. $1(7) = 7 = a$ $1(-17) = -17 = c$
                 $-119 + 1 = -118 = b$ The answers are incorrect.
The correct answers are 17 and $-\frac{1}{7}$

b. When the answers are two fractions with radical form.

Example 5. Check the answers $\frac{2+\sqrt{3}}{3}$ and $\frac{2-\sqrt{3}}{3}$ of the quadratic equation $9x^2 - 12x + 1 = 0$

Solution. $3(3) = 9 = a$ $(2+\sqrt{3})(2-\sqrt{3}) = 4 - 3 = 1 = c$
Diagonal sum $3(2+\sqrt{3}) + 3(2-\sqrt{3}) = 6 + 3\sqrt{3} + 6 - 3\sqrt{3} = 12 = -b$
The answers are correct.

Example 6. Check the answers $\dfrac{-2+\sqrt{5}}{3}$, $\dfrac{-2-\sqrt{5}}{3}$ of the quadratic equation $9x^2 + 12x - 1 = 0$

Solution. $3(3) = 9 = a$ $(-2+\sqrt{5})(-2-\sqrt{5}) = 4 - 5 = -1 = c$
Diagonal sum $-6 - 6 = -12 = -b$ The answers are correct.

c. When the answers are two conjugate complex numbers.

Example 7. Check the answers $\dfrac{-3+4i}{2}$, $\dfrac{-3-4i}{2}$ of the quadratic equation $4x^2 + 12x + 25 = 0$

Solution. $2(2) = 4 = a$ $(-3+4i)(-3-4i) = 9 + 16 = 25 = c$
$-6 - 6 = -12 = -b$ The answers are correct.

Example 8. Check the answers $\dfrac{2+5i}{4}$, $\dfrac{2-5i}{4}$ of the quadratic equation $16x^2 - 16x + 9 = 0$

Solution. $4(4) = 16 = a$ $(2+5i)(2-5i) = 4 + 25 = 29 \neq c$
$8 + 8 = 16 = -b$ The answers are incorrect
The correct answers are $\dfrac{2 \pm i\sqrt{5}}{4}$

Exercises on checking answers when solving quadratic equations.

1. $35x^2 + 61x - 24 = 0$ $(-\dfrac{3}{5})$, $(-\dfrac{8}{7})$
2. $15x^2 + x - 40 = 0$ $(-\dfrac{5}{3})$, $(\dfrac{8}{5})$
3. $6x^2 + x - 35 = 0$ $(-\dfrac{5}{2})$, $(\dfrac{7}{3})$
4. $-4x^2 + 39x - 56 = 0$ $(\dfrac{7}{4})$, (8)
5. $3x^2 + 2x - 8 = 0$ $(-\dfrac{4}{3})$, (-2)
6. $21x^2 - 5x - 6 = 0$ $(-\dfrac{2}{3})$, $(\dfrac{3}{7})$
7. $49x^2 - 70x + 19 = 0$ $(\dfrac{5 \pm \sqrt{6}}{7})$
8. $25x^2 - 20x + 1 = 0$ $(\dfrac{2 \pm \sqrt{3}}{5})$
9. $9x^2 - 30x + 19 = 0$ $(\dfrac{5 \pm \sqrt{6}}{3})$
10. $49x^2 + 42x + 4 = 0$ $(\dfrac{3 \pm \sqrt{5}}{7})$
11. $16x^2 - 16x + 13 = 0$ $(\dfrac{-2 \pm 3i}{4})$
12. $25x^2 - 30x + 13 = 0$ $(\dfrac{3 \pm 5i}{5})$
13. $25x^2 - 20x + 7 = 0$ $(\dfrac{-2 \pm 3i}{5})$
14. $4x^2 + 20x + 34 = 0$ $(\dfrac{5 \pm 3i}{2})$

1-8. CHAPTER REVIEW AND REVIEW EXERCISES.

SOLVING QUADRATIC EQUATIONS $ax^2 + bx + c = 0$

The standard form of a quadratic equation is $ax^2 + bx + c = 0$ where a is not equal to zero.

The equation has a solution set with at maximum two **real roots** or two values of x that make the equation true. Depending on the values of the constants a, b, and c, there may be one **double root** or no real root at all.

There are a few common methods to solve a quadratic equation. The general method uses the quadratic formulas. When b is even, there are simplified quadratic formulas. Other methods use mental math combined with trial and error to quickly find the real roots.

Solving by the Quadratic Formulas.

a. <u>General formulas</u>

The Discriminant: $D = b^2 - 4ac$

If $D > 0$ There are two real roots given by formulas

$$x = \frac{-b \pm \sqrt{D}}{2a}$$

If $D = 0$ There is one double real root, $x_1 = x_2 = -\frac{b}{2a}$

If $D < 0$ There are no real roots.
                    Note. However, there are two roots in the form of two conjugate complex numbers that we will learn later.

The **Sum** of the roots $S = x_1 + x_2 = -\frac{b}{a}$

The **Product** of the roots $P = x_1 \cdot x_2 = \frac{c}{a}$

b. <u>Simplified formulas when b is even number (b = 2b')</u>

We particularly recommend students use these simplified formulas because they apply to 50% of the cases.

The Discriminant $D' = b'^2 - ac$

If $D' > 0$ There are two real roots
$$x = \frac{-b' \pm \sqrt{D'}}{a}$$

If $D' = 0$ There is one double real root, $x_1 = x_2 = -\frac{b'}{a}$

If $D' < 0$ There is no real roots.

Solving by factoring

This method uses mental math combined with trial and error to factor the left side of the equation into two binomials of x. After the equation is factored, the roots will be obtained by solving the two binomials for x. This method usually requires experience and a sharp mental math ability. In general, there are many permutations involved, so it is hard to mentally match the terms.

Solving by the Diagonal Sum Method

This method has the advantage of directly obtaining the real roots in the form of two fractions. The concept of the method is to find two fractions whose sum is $(-\frac{b}{a})$ and whose product is $(\frac{c}{a})$.

Based on the values of the constant a and c, students first select two root sets, each one is in the form of two fractions. The sign selection is based on the rule of sign for real roots of a quadratic equation. Then a mental computation of the **diagonal sum** of all the fractions inside the two sets will indicate which set is the solution set.

The rule of sign for real roots helps reduce the number of permutations considerably. Strictly following the operation steps and through experience, students can find the correct combinations easily.

Review exercises on solving quadratic equations

1. $3x^2 + 4x - 4 = 0$
2. $3x^2 - 10x - 13 = 0$
3. $2x^2 + 7x - 4 = 0$
4. $5x^2 + 9x - 2 = 0$
5. $2x^2 - 9x + 9 = 0$
6. $3x^2 - 8x - 11 = 0$
7. $15x^2 - 43x + 22 = 0$
8. $19x^2 - 18x - 1 = 0$
9. $7x^2 - 3x - 10 = 0$
10. $12x^2 + 69x - 18 = 0$
11. $21x^2 + 26x - 15 = 0$
12. $15x^2 - 4x - 35 = 0$
13. $15x^2 + x - 28 = 0$
14. $20x^2 - 31x + 12 = 0$
15. $35x^2 + 24x - 27 = 0$
16. $-12x^2 + 43x - 35 = 0$
17. $5x^2 + 64x - 13 = 0$
18. $7x^2 + 76x - 11 = 0$
19. $8x^2 - 38x + 17 = 0$
20. $6x^2 + 41x + 13 = 0$
21. $9x^2 - 12x + 1 = 0$
22. $4x^2 - 6x + 1 = 0$
23. $3x^2 - 4x - 1 = 0$
24. $25x^2 - 20x - 1 = 0$
25. $9x^2 - 30x + 19 = 0$
26. $49x^2 + 42x + 4 = 0$

CHAPTER 2

SOLVING QUADRATIC EQUATIONS – SPECIAL CASES

The quadratic equation in the general form, in which a, b, c are real numbers, is

$$ax^2 + bx + c = 0 \qquad (a \neq 0) \qquad (1)$$

There are a few special cases of quadratic equations students need to learn how to solve.

2-1. SOLVING THE QUADRATIC EQUATION $x^2 + bx + c = 0$

When the constant a = 1, the quadratic equation is simplified to the form, in which b, c are real numbers:

$$x^2 + bx + c = 0 \qquad (1)$$

Solving by the quadratic formulas, as described in Paragraph 1-3, gets easier since the formulas are simplified.

In general, solving by the quadratic formulas is always necessary when the Discriminant D (or D') is not a **perfect square** or when the square root of D (or D') is an **irrational number**.

In other cases, the mental math approach can get the answer faster than using the quadratic formulas.

When the constant a = 1, solving a quadratic equation becomes solving a popular puzzle: **find two numbers knowing their sum and their product**. The approach uses mental math and requires some practice and experience.

Students can use the Diagonal Sum method that makes the puzzle solving easier by setting up practical steps to follow.

Solving by factoring may take longer in the process of factoring and solving the two binomials of x.

Another method of solving a quadratic equation, the method of completing the square, deserves to be studied.

2-1-1. Solving by the quadratic formulas

When $a = 1$, the quadratic formulas become simplified.

Case 1. When b is an odd number

The Discriminant is $\quad D = b^2 - 4c$

The two real roots are $\quad \dfrac{-b \pm \sqrt{D}}{2}$

Sum of the roots: $S = x_1 + x_2 = (-b)$

Product of roots: $P = x_1 \cdot x_2 = (c)$

Example 1. Solve the equation $\quad x^2 - 5x + 5 = 0$

Solution. Discriminant $D = 25 - 20 = 5$

The two roots are $\dfrac{5 \pm \sqrt{5}}{2}$

Example 2. Solve the equation $\quad x^2 - 7x + 10 = 0$

Solution. Discriminant $= 49 - 40 = 9$

The roots are $\dfrac{7 \pm 3}{2} \quad$ or $\quad$ (5) and (2)

Case 2. When b is an even number ($b = 2b'$ or $b' = \dfrac{b}{2}$)

The Discriminant $\quad D' = b'^2 - c$

Roots $\quad (-b' + \sqrt{D'}) \quad (-b' - \sqrt{D'})$

Sum of roots: $\quad S = x_1 + x_2 = -b = -2b'$

Product of roots: $\quad P = x_1 \cdot x_2 = (c)$

Example 1. Solve the equation $\quad x^2 - 4x + 2 = 0$

Solution. Discriminant $\quad D' = 4 - 2 = 2$

The two roots are $\quad 2 \pm \sqrt{2}$

Example 2. Solve $\quad x^2 + 6x + 5 = 0$

Solution. Discriminant $\quad D' = 9 - 5 = 4$

Two roots are $\quad -3 \pm 2 \quad$ or $\quad (-1)$ and (-5)

Exercises on solving by quadratic formulas

1. $x^2 - 3x + 1 = 0$
2. $-x^2 + 4x + 2 = 0$
3. $x^2 - 5x + 5 = 0$
4. $x^2 - 12x + 28 = 0$
5. $x^2 + 3x - 1 = 0$
6. $-x^2 + 7x - 17 = 0$
7. $x^2 - 15x + 56 = 0$
8. $x^2 + 17x + 72 = 0$
9. $x^2 - 10x - 39 = 0$
10. $x^2 - 10x - 56 = 0$

2-1-2. Quick solving by mental math.

In many cases, solving the equation $x^2 + bx + c = 0$ by mental math is faster than using the quadratic formulas. The purpose of the mental math approach is to solve a popular puzzle : find two numbers knowing their sum and their product. There are three mental math approaches.

The **first approach** is purely mental. Students find the solution by guessing two numbers whose sum is $-b$ and whose product is c.

Example 1. Solve the equation $\quad x^2 + 5x + 6 = 0$.

Solution. Find by guessing two numbers whose sum is (-5) and whose product is (6). They are (-2) and (-3).

There are no specific instructions for guessing. This approach requires students to have a sharp mental math ability and some experience.

If students do not feel comfortable with this purely mental approach, they can use the **Diagonal Sum Method** to solve the puzzle quickly. This method reduces the number of permutations involved by using the rule of sign. It also sets up practical operation steps to follow. This method is particularly convenient when b, c are big numbers.

The **third approach**, the solving by factoring method (see Paragraph 1-5), may takes longer in factoring and solving the two binomials.

2-1-3. Solving by the Diagonal Sum method.

When $a = 1$, the diagonal sum $c_1 a_2 + c_2 a_1$ becomes the sum of the two factors $(c_1 + c_2)$ of c (See Paragraph 1-5). The operation steps of the Diagonal Sum Method become the following:

Step1. First, use the rule of sign to know in advance the signs of the two real roots.
Step 2. Write all the possible combinations of factors of c.
Step 3. Mentally compute the sum of each combination and compare it to (-b). If it matches, the two related factors are the real roots. If the sum in one combination is equal to (b), the real roots have opposite signs to the ones in the combination.

Example 1. Solve $\quad x^2 - 9x + 14 = 0$

Solution. Both roots are positive. Write all the combinations of factors of 14.

$\quad\quad$ (2 , 7) $\quad\quad\quad\quad\quad$ (1 , 14).

Mentally compute their **sum** and compare it to either **(-b) or (b)**

$2 + 7 = 9 = -\mathbf{b}$ $\quad\quad\quad\quad$ $(1 + 14 = 15)$

The first set of roots is the answer since its sum is equal to (-b). The real roots are 2 and 7.

Example 2. Solve $x^2 + 27x + 50 = 0$

Solution. Both roots are negative. Write all combinations of factors of 50, and mentally compute their sum and then compare it to either (-b) or (b)

(-5 , -10)   (-2 , -25)   (-1 , - 50)
                  (-27 = -b)

The real roots are -2 and -25

Example 3. Solve $x^2 + 34x - 72 = 0$

Solution. The two roots are opposite signs. Write all the factor combination of -72 and then mentally compute their sum

(-8 , 9) (8 , -9)   (-2 , 36) (2 , -36)   (-4 , 18) (4 , -18)   (-1 , 72) (1 , -72)
  (1)   (-1)   (34)   **(-34)**   (14)   (-14)   (71)   (-71)

The real roots are 2 and -36

Important Remark. We see that there are two opposite sets in each combination. As a consequence, we only have to write and compute one sum for each combination. The other sum can be obtained by taking the opposite.

Example 4. Solve $x^2 + 14x - 72 = 0$

Solution. The two roots are opposite sign. Write only one set of roots for each combination. Mentally compute the sum and compare it with either
(b) or (-b). Stop doing it once you get the sum equal to either (b) or (-b).

(-8 , 9)   (-4 , 18)   (-2 , 36)   (-1 , 72)
9 - 8 = 1   18 – 4 = 14 = **b**

The real roots (4 , - 18) are **opposite** to the ones of the second set.

33

Exercises on solving by the Diagonal Sum method.

1. $x^2 + 11x + 24 = 0$
2. $x^2 + 10x + 25 = 0$
3. $x^2 - 8x + 12 = 0$
4. $x^2 + 4x - 21 = 0$
5. $x^2 - 9x + 14 = 0$
6. $x^2 + x - 20 = 0$
7. $x^2 + x - 6 = 0$
8. $x^2 - 2x - 15 = 0$
9. $x^2 + x - 72 = 0$
10. $x^2 + 16x - 36 = 0$
11. $x^2 + 22x - 48 = 0$
12. $x^2 + 8x - 48 = 0$
13. $-x^2 - 26x + 56 = 0$
14. $x^2 + 12x - 56 = 0$
15. $-x^2 + 16x - 48 = 0$
16. $x^2 + 26x + 48 = 0$
17. $x^2 + 34x - 72 = 0$
18. $-x^2 + 17x - 72 = 0$
19. $x^2 - 24x + 23 = 0$
20. $x^2 - 16x - 17 = 0$

2-2. SOLVING BY COMPLETING THE SQUARE.

Equation to solve $\quad x^2 + bx + c = 0 \quad$ (1)

This method transforms the left side of equation (1) into the square of a binomial by using common identities of binomials. These common identities are:

$(a - b)^2 = a^2 - 2ab + b^2$
$(a + b)^2 = a^2 + 2ab + b^2$

<u>Example 1.</u> Solve by completing the square $\quad x^2 + 4x - 5 = 0$

Solution.

$x^2 + 4x = 5$ Moving term

$x^2 + 4x + 4 = 5 + 4$ Add (4) to both sides to make the left side a perfect square

$(x + 2)^2 = 9$ Binomial identity

$(x + 2) = \pm 3$ Take the square root of both sides

$x = -2 \pm 3$ Answers

The two real roots are $\quad -2 + 3 = 1 \quad$ and $\quad -2 - 3 = -5$

Example 2. Solve by completing the square $x^2 - 8x + 15 = 0$

$$x^2 - 8x = -15 \qquad \text{Moving terms}$$

$$x^2 - 8x + 16 = -15 + 16 \qquad \text{Add 16 to both sides to make the left side a perfect square}$$

$$(x - 4)^2 = 1 \qquad \text{Binomial identity}$$

$$x - 4 = \pm 1 \qquad \text{Square root of both sides}$$

The two real roots are 5 and 3   Answers

Exercises on solving by completing the square.

1. $x^2 - 4x - 5 = 0$ 2. $x^2 + x - 6 = 0$

3. $x^2 - 2x - 15 = 0$ 4. $x^2 + 11x + 24 = 0$

5. $x^2 + 4x - 21 = 0$ 6. $x^2 - 9x + 14 = 0$

7. $x^2 - 8x + 12 = 0$ 8. $x^2 - 6x + 8 = 0$

9. $x^2 + 10x + 16 = 0$ 10. $x^2 - 12x - 13 = 0$

2-3. QUADRATIC EQUATION WITH PARAMETERS.

Sometimes the constants a, b, and c are expressed in terms of a parameter m. Depending on the value of m, these constants have different values creating various situations in which the quadratic equation possesses some specific characteristics. Discussion of these situations when m varies is the main goal when solving equations with parameters.

Example 1   Given the equation $2x^2 + 4x + m = 0$
For what values of m does this equation have two real roots?

Solution. This equation has two real roots when the discriminant is positive.

$$D = 16 - 8m > 0 \qquad \text{Condition to have two real roots}$$
$$8m > 16 \qquad \text{Moving terms.}$$
$$m > 2 \qquad \text{Answer}$$

Example 2. Given the equation $x^2 + 3x + m = 0$

Find the value of m so that one real root is twice the other.

Solution

$x_1 = 2x_2$	Given data
$2x_2 + x_2 = -3$	Sum of the two roots equal to -b
$3x_2 = -3$	Combine like terms
$x_2 = -1$ (1)	First value of x
$x_1 \cdot x_2 = 2x_2^2 = m$ (2)	Product of roots. Second value of x
$2(-1)^2 = m$	Compare two values (1) and (2)
$m = 2$	Answer

Check by replacing $m = 2$ in the equation to solve $x^2 + 3x + 2 = 0$

There are two real roots (-1) and (-2), one is twice the other.

Example 3. Equation $7x^2 + (3 - t)x - 2(t - 5) = 0$ (1)

Find the value of t so that the equation has (-1) as one of its real roots.

Solution. The equation has one real root equal to (-1) when $a - b + c = 0$

$7 - 3 + t - 2t + 10 = 0$	Tip 2 (See Paragraph 1-5)
$t = 14$	Combine like terms. Answer

Check

$7x^2 - 11x - 18 = 0$	Replace $t = 14$ in equation (1)
Roots (-1) and ($\frac{18}{7}$)	Solving quadratic equation.

Example 4. Find m in this equation so that the two real roots are reciprocals of each other.

$$(m + 1)x^2 - 10x - 3m + 9 = 0 \qquad (1)$$

Solution. The two real roots are reciprocals when $\dfrac{c}{a} = 1$ or $a = c$

$m + 1 = -3m + 9$ Roots are reciprocal $a = c$
$4m = 8$
$m = 2$ Answer

Check: $3x^2 - 10x + 3 = 0$ Replace $m = 2$ in (1)

Two reciprocal roots $(\dfrac{1}{3})$ and (3)   Solve by Diagonal Sum method

Exercises on solving quadratic equation with parameters.

Find m so that these equations have double roots:

1. $4x^2 + 2mx + 1 = 0$ 2. $4x^2 - 2(m + 3)x - m = 0$

Find m in these equations so that the difference of two real roots is equal to 2

3. $x^2 - mx + 1 = 0$ 4. $x^2 + 4x - m = 0$

Find m in these equations so that each one has a real root equal to 5

5. $2x^2 + mx - 10 = 0$ 6. $mx^2 - 9x - 5 = 0$

7. Find m so that the equation $x^2 - 4x + m = 0$ has the sum of the square of its roots equal to 26. (Hint: $x_1^2 + x_2^2 = S^2 - 2P = 26$)

8. Given the equation $x^2 - 6x - m^2 = 0$. Prove that the equation has two real roots x and x which are opposite sign for any value of m. Find m so that the two roots are related by this relation $x_1 + 3x_2 = 4$.

9. Given the equation $x^2 + px + q = 0$. Find p and q so that the equation has two real roots (p) and (q).

10. Find m so that the two real roots are reciprocals $mx^2 - 17x - m + 8 = 0$

2-4. SOLVING BIQUADRATIC EQUATION

This equation is in the form $\quad ax^4 + bx^2 + c = 0 \quad (1)$

Let $X = x^2$, then the equation becomes a regular quadratic equation,

$$aX^2 + bX + c = 0 \quad (2)$$

The values of the two roots (X_1) and (X_2) are given by common methods of solving quadratic equations. Then find the roots of equation (1) by solving $x^2 = X_1$, and $x^2 = X_2$

Remark. Only the positive real root(s) of the quadratic equation (2) can give the real roots of the biquadratic equation (1)

<u>Example 1.</u>   Solve the equation $\quad x^4 - 5x^2 + 4 = 0$

$X^2 - 5X + 4 = 0$ Let $X = x^2$

Two roots: $(X_1 = 1)\;\;(X_2 = 4)$ Tip case 1 $(a + b + c = 0)$

Four real roots: $\pm 1 \quad$ and $\quad \pm 2$ Solve for x by the equation $x^2 = X$

<u>Example 2.</u>   Given a biquadratic equation $ax^4 + bx^2 + c = 0 \quad (1)$

Prove that
a. If this equation has four roots, then their sum is always equal to zero and their product is $\dfrac{c}{a}$.

b. If $ac < 0$, meaning a and c are opposite signs, then this equation only has two opposite real roots.

<u>Solution.</u>   Let $x^2 = X$ and transform (1) to $aX^2 + bX + c = 0 \quad (2)$
This equation (2) has two roots: (X_1) and (X_2)

  a. If the biquadratic equation (1) has 4 roots, it means there are 2 sets of opposite roots, $(\pm\sqrt{X_1})$ and $(\pm\sqrt{X_2})$. Obviously, their sum is always zero

Their product is $(+\sqrt{X_1})(-\sqrt{X_1})(+\sqrt{X_2})(-\sqrt{X_2}) = X_1 \cdot X_2 = \dfrac{c}{a}$

  b. If $ac < 0$, the two roots X_1 and X_2 of equation (2) are of opposite signs. Suppose X_1 is the positive root. The negative root X_2 does not give real roots. The unique positive real root X_1 gives two opposite real roots $(\pm\sqrt{X_1})$

Exercises on biquadratic equations.

1. $x^4 - 7x^2 + 6 = 0$
2. $2x^4 + 5x^2 + 2 = 0$
3. $x^4 - 4 = 0$
4. $x^4 - 5 = 2(x^2 - 1)$
5. $3x^4 + 5x^2 + 2 = 0$
6. $8x^4 - 22x^2 + 15 = 0$
7. $6x^4 + x^2 - 12 = 0$
8. $2x^4 - 9x^2 + 4 = 0$
9. $-6x^4 + 7x^2 + 20 = 0$
10. $6x^4 - 11x^2 - 35 = 0$
11. $20x^4 - 31x^2 + 12 = 0$
12. $35x^4 + 24x^2 - 27 = 0$
13. $15x^4 + 10x^2 - 40 = 0$
14. $-4x^4 + 39x^2 - 56 = 0$

2-5. EQUATIONS WITH RATIONAL EXPRESSIONS.

These equations have rational expressions in x in their terms. To solve these equations, transform them into quadratic equations by using common transformations.

<u>Example 1</u>. Solve $\quad \dfrac{1}{x+1} = 2x$

Solution. The equation is undefined when $x = -1$ and $x = 0$

If $x \neq -1, x \neq 0,\quad 1 = 2x^2 + 2x \qquad$ Multiply both side by $(x+1)$
$\qquad\qquad\qquad 2x^2 + 2x - 1 = 0 \qquad$ Simplify and moving term

Discriminant $D' = 1 + 2 = 3$

Two real roots $\quad x = \dfrac{-1 \pm \sqrt{3}}{2} \qquad$ Answers.

<u>Example 2</u>. Solve $\quad \dfrac{2x-1}{x+2} = x - 2$

Solution. The equation is undefined when $x = \pm 2$

If $x \neq \pm 2 \qquad 2x - 1 = x^2 - 4 \qquad$ Cross multiplication
$\qquad\qquad\qquad x^2 - 2x - 3 = 0 \qquad$ Simplify and move terms
$\qquad$ Roots $\qquad (-1)\ (3) \qquad$ Answers.

Exercises on solving equations with rational expressions.

1. $\dfrac{-2(11x+6)}{2x-3} = 3x+1$
2. $\dfrac{x(5x+11)}{x+1} = 2$
3. $\dfrac{3x^2+11x-9}{x+4} = 1$
4. $x+3 = \dfrac{6x^2+8x-14}{x-4}$
5. $\dfrac{3x^2+13x+5}{x+3} = x+3$
6. $\dfrac{3x^2+11x-9}{3x+2} = x-1$
7. $\dfrac{3x-1}{2x+3} = 4x$
8. $\dfrac{2x-1}{3x+2} = x+4$
9. $3x-1 = \dfrac{2(x+1)}{2x+5}$
10. $\dfrac{7-5x}{(x-3)(2x+1)} = 3$
11. $\dfrac{x^2+64x-49}{x+1} = 16(x-1)$
12. $4x+1 = \dfrac{2(9x+5)}{4x-5}$
13. $\dfrac{2x-5}{x+3} = \dfrac{x+3}{2x-5}$
14. $\dfrac{15x^2+4x-19}{x+3} = 3$

2-6. EQUATIONS WITH RADICAL EXPRESSIONS

Some terms of an equation may have radical expressions. To transform the radical term, isolate it on one side of the equation and then square both sides of the equation. Solve the resulting quadratic equation by using common solving methods.

In this paragraph, we only cover simple equations with radical expressions. After the solution set is found, we can check to see if it is justified.

Note: There are always conditions which applied to the variable x so that the equation makes sense

<u>Example 1</u>. Solve equation $\sqrt{2x-3} = x-3$ \qquad (1)

Solution
Condition $2x-3 \geq 0$ \quad (1) \qquad Radical number property
and \qquad\qquad $x \geq 3$ \quad (2) \qquad Right side should be positive

Condition \quad $x \geq 3$ \qquad\qquad Combining the 2 conditions

$$2x-3 = (x-3)^2 \qquad \text{Square both sides}$$
$$2x-3 = x^2-6x+9 \qquad \text{Binomial identity}$$
$$x^2-8x+12 = 0 \qquad \text{Like terms and moving term}$$

Two real roots   (2) and (6)   Solving quadratic equation

    Root x = 2   Not justified since less than 3
    Root x = 6   Accepted since greater than 3

Check: Replace x = 6 in equation (1) . It is true.

Example 2. Solve equation $\sqrt{3x-2}$ - x + 2 = 0

Solution. $\sqrt{3x-2}$ = x - 2   Moving terms

Conditions   x – 2 ≥ 0   (1)   Right side positive
              3x – 2 ≥ 0   (2)   Radical number property

Condition   x ≥ 2   Combine two conditions

      3x – 2 = (x – 2)²   Square both sides
      x² - 7x + 6 = 0   Like terms and moving terms

Two real roots   (1) and (6)   Solving quadratic equation

    Root x = 1   Rejected since less than 2
    Root x = 6   Accepted since greater than 2

Check by replacing x = 6 in equation (1). It is true.

Example 3. Solve $\sqrt{3x(x-2)}$ - (x + 4) = 0 (1)

Solution. $\sqrt{3x(x-2)}$ = x + 4   Moving terms

Condition: x + 4 ≥ 0   Right side positive

or   x ≥ -4   Condition (1)

Condition: 3x (x –2) ≥ 0   Radical number property.

    x = 0 and   x = 2   Two real roots

    x ≤ 0   or   x ≥ 2   Condition (2). Rule of sign for a
                                                    trinomial. See Paragraph 3-3

$\quad -4 \leq x \leq 0 \quad$ or $\quad x \geq 2 \qquad$ Combination of 2 conditions

$\quad\quad 3x(x-2) = (x+4)^2 \qquad$ Square both sides

$\quad\quad 3x^2 - 6x = x^2 + 8x + 16 \qquad$ Binomial identity

$\quad\quad 2x^2 - 14x - 16 = 0 \qquad$ Moving and combining like terms

Two real roots $\;-1\;$ and $\;8 \qquad$ Solving quadratic equation

Root $\;x = -1 \qquad$ Accepted since $\;-4 < -1 < 0$

Root $\;x = 8 \qquad$ Accepted since $\;8 > 2$

Check by replacing $x = -1$ and $x = 8$ in equation (1). It is true.

Example 4. Solve $\quad 3x + 1 - \sqrt{x^2 + 8x + 7} = 0 \qquad (1)$

Solution. $\quad 3x + 1 = \sqrt{x^2 + 8x + 7} \qquad$ Moving terms

Condition $\;x^2 + 8x + 7 \geq 0 \qquad$ Radical number property

Two real roots $\;-7\;$ and $\;-1 \qquad$ Solving quadratic equation

$\quad x \leq -7 \;$ or $\; x \geq -1 \qquad (1) \qquad$ Rule of sign. See 3-3

Condition $\;3x + 1 \geq 0 \qquad (2) \qquad$ Left side should be positive

$\quad x \leq -7 \;$ or $\; -\dfrac{1}{3} \leq x \qquad$ Combine 2 conditions

$\quad 9x^2 + 6x + 1 = x^2 + 8x + 7 \qquad$ Square both sides

$\quad 8x^2 - 2x - 6 = 0 \qquad$ Moving and combining terms

Two real roots $\;(1)$ and $(-\dfrac{3}{4}) \qquad$ Solving quadratic equation

Root $\;x = 1 \qquad$ **Accepted** since greater than $-\dfrac{1}{3}$

Root $\;x = -\dfrac{3}{4} \qquad$ Not accepted since greater than -7

Students may check by replacing $x = 1$ into equation (1)

Exercises on solving equations with radical expressions. Solve

1. $x - 4 - \sqrt{2x - 5} = 0$
2. $\sqrt{2x - 6} - x + 3 = 0$
3. $\sqrt{5x + 10} = 8 - x$
4. $x - 3 = \sqrt{2x - 3}$
5. $\sqrt{3x^2 - 7x + 4} = 3x - 4$
6. $2x - 5 - \sqrt{2x^2 - 7x + 5} = 0$
7. $\sqrt{2x^2 - 4x - 5} = 2x - 10$
8. $\sqrt{x^2 - 5x + 4} - 3x + 12 = 0$
9. $\sqrt{2x^2 - 7x - 9} = 2x - 9$
10. $3x - 10 - \sqrt{3x^2 - 7x - 10} = 0$

2-7. CHAPTER REVIEW: Solving quadratic equations– Special cases

Solving the quadratic equation $x^2 + bx + c = 0$

In the particular case when $a = 1$, the quadratic equation is simplified to the form, in which b, c are real numbers: $x^2 + bx + c = 0$

Solving this equation using the quadratic formulas, as described in Chapter 1, gets easier since the formulas are simplified. In general solving by quadratic formulas is necessary when the Discriminant D (or D') is not a **perfect square** or when the square root of D (or D') is an **irrational** number.
In other cases, the solving by mental math can get the answer faster than using the quadratic formulas.
When $a = 1$, solving a quadratic equation becomes solving a popular puzzle: find two numbers knowing their sum and their product. There are two main mental math approaches for solving a quadratic equation quickly.
The first approach, the Diagonal Sum method, makes solving the puzzle easier by setting up practical operation steps for students to follow.
The second approach, solving by factoring, may take longer in the factoring process and then in solving the two binomials
Another method, solving by completing the square, deserves to be looked at.

Solving quadratic equation with parameters.

Sometimes the constants of the quadratic equations are expressed in term of a parameter m. When m varies, the constants get different values creating various situations. Discussing these situations when m varies is the main goal in solving equations with parameter.

Example. Find m so that the quadratic equation $2x^2 + 3x - m = 0$ has a double root.

Solving equations with rational expressions

Some quadratic equations have rational expressions in their terms. In order to solve these types of equations, transform them into quadratic equations by using common transformations. The most common transformations use the least common denominator (LCD) or cross multiplication.

Example of equation with rational expression. $\dfrac{ax + b}{cx + a} = ex + f$

Solving biquadratic equations

The biquadratic equation is in the standard form $x^4 + bx^2 + c = 0$ (1)

Replace $X = x^2$ in the equation (1) to get a quadratic equation in term of X:

$$X^2 + bX + c = 0 \quad (2)$$

Find the two roots (X_1) and (X_2) of the equation (2) by common methods to solve a quadratic equation. Then, solve for x by the relation $x^2 = X$.
Note that only positive values of X can give the real values of x.

Solving quadratic equations with radical expressions.

To eliminate the radical term, isolate it on one side of the equation. Square both sides of the equation then solve by common solving methods. Apply conditions to variable x so that the equation makes sense. These conditions sometimes make the solving complicated.
You can check the solution set to see if it is justified.

2-8. REVIEW EXERSICES.

a. Solving by quadratic formulas (b odd number).

1. $x^2 - 7x + 12 = 0$
2. $x^2 - 9x + 20 = 0$
3. $x^2 - 3x + 1 = 0$
4. $x^2 + 7x - 11 = 0$
5. $x^2 - 3x - 1 = 0$
6. $x^2 - 5x + 5 = 0$
7. $x^2 - 11x + 18 = 0$
8. $x^2 - 19x + 34 = 0$
9. $-x^2 + 5x - 4 = 0$
10. $x^2 - 13x + 22 = 0$

b. Solving by quadratic formulas (b even number).

1. $x^2 + 4x - 4 = 0$
2. $x^2 - 6x + 8 = 0$
3. $x^2 + 8x - 4 = 0$
4. $x^2 + 12x + 24 = 0$
5. $x^2 - 4x - 14 = 0$
6. $x^2 + 6x - 3 = 0$
7. $x^2 + 8x - 16 = 0$
8. $x^2 - 10x + 13 = 0$
9. $x^2 - 28x + 52 = 0$
10. $x^2 - 20x + 80 = 0$

c. Solving by completing the square.

1. $x^2 - 6x + 5 = 0$
2. $x^2 + 12x + 11 = 0$
3. $x^2 - 4x - 9 = 0$
4. $x^2 - 5x + 3 = 0$
5. $x^2 - 7x + 10 = 0$
6. $x^2 + 8x + 7 = 0$
7. $x^2 + 11x + 10 = 0$
8. $x^2 + 9x + 16 = 0$

d. Solving quadratic equations with parameter.

Find m so that the equation has one root equal to $(-\frac{c}{a})$

1. $3x^2 - 13x + m - 17 = 0$
2. $2mx^2 - 11x - 9 - m = 0$

Find m so that the two real roots are reciprocals.

3. $11x^2 - 122x + m - 2 = 0$
4. $-6x^2 + 37x - \frac{3}{m} = 0$

Find m so that one real root of each equation is 3

5. $mx^2 - 7x - 2m = 0$
6. $5x^2 - (m + 1)x - 3m = 0$

Find m so that these equations have one real root which is three times of the other.

7. $mx^2 - 12x + 23 + 3 = 0$
8. $x^2 - 4mx + 5m + 2 = 0$

e. Solving biquadratic equations.

1. $3x^4 - 10x^2 + 7 = 0$
2. $5x^4 + 54x^2 - 11 = 0$
3. $6x^4 - 29x^2 + 13 = 0$
4. $5x^4 + 12x^2 + 5 = 0$
5. $7x^4 - 25x^2 - 12 = 0$
6. $8x^4 - 14x^2 - 15 = 0$
7. $6x^4 - 23x^2 + 7 = 0$
8. $x^4 - 7x^2 + 10 = 0$
9. $x^4 - 10x^2 + 21 = 0$
10. $x^4 + 11x^2 - 60 = 0$

f. Solving equations with rational expressions.

1. $\dfrac{7x^2 + 21x - 14}{x - 2} = x + 2$

2. $\dfrac{3x + 16}{3x - 2} = 3x + 2$

3. $x + 2 = \dfrac{8x^2 - 4x - 8}{x + 2}$

4. $\dfrac{2x + 1}{x^2 + 15x - 19} = \dfrac{1}{x - 3}$

5. $\dfrac{2}{x - 4} = \dfrac{x - 5}{x^2 - 5x - 5}$

6. $\dfrac{3}{x - 2} = \dfrac{x + 2}{x^2 - 4x - 12}$

7. $\dfrac{2x + 5}{x^2 - 10x - 95} = \dfrac{1}{x + 3}$

8. $\dfrac{x^2 + 5x + 11}{4x - 3} = \dfrac{x - 3}{2}$

9. $\dfrac{x + 7}{5} = \dfrac{3x^2 - 4x - 5}{x - 4}$

10. $\dfrac{5x^2 + 3x - 6}{x + 3} = \dfrac{x + 3}{3}$

g. Solving equations with radical expression.

1. $\sqrt{2x - 5} = x - 4$

2. $\sqrt{3x + 1} = 9 - x$

3. $11 - 2x = \sqrt{3x + 16}$

4. $17 - 3x = \sqrt{6x + 1}$

5. $7x - 9 = \sqrt{31 - 3x}$

6. $\sqrt{3x^2 + 4x - 15} = x + 3$

7. $\sqrt{4x^2 - 7x - 11} - (x + 1) = 0$

8. $\sqrt{-3x^2 - 4x + 7} = x - 1$

9. $\sqrt{x^2 - 7x + 10} = 2x - 10$

10. $\sqrt{x^2 - 5x + 4} - 12 + 2x = 0$

11. $\sqrt{3x^2 - 10x + 3} = 2x - 6$

12. $x + 3 = \sqrt{3x^2 - 2x - 1}$

CHAPTER 3

SOLVING QUADRATIC INEQUALITIES AND SYSTEM OF QUADRATIC INEQUALITIES IN ONE VARIABLE

3-1. GENERALITIES ON SOLVING QUADRATIC INEQUALITIES

3-1-1. Standard form of a quadratic inequality.

The standard form of a quadratic inequality is

$$f(x) = ax^2 + bx + c > 0 \quad \text{or} \quad f(x) = ax^2 + bx + c < 0 \quad (1)$$

in which $f(x)$ is a trinomial expression of variable x, and a, b, c, are real numbers with a not equal to zero. If $f(x)$ is shown in another form, transform it to the standard form using common transformations such as factoring, moving and combining like terms. Solving a quadratic inequality means finding the solution set whose values of x make the inequality true.

3-1-2. Solving approaches.

There are many approaches to solving quadratic inequalities. We cover here the two most common approaches.

1. Approach 1 uses the 2 roots of the quadratic equation $f(x) = 0$ and the number line. Suppose x_1 and x_2 are the two roots of $f(x) = 0$. The graph of them divides the number line into three parts, 2 rays and one segment. To find the solution set, replace the coordinate zero of the origin into the inequality. If it makes the inequality true, then the origin is on the true part(s) that belongs to the solution set. If not, the origin is outside the solution set.

2. Approach 2 is based on the application of a theorem that shows the sign change (from negative to positive, or vice versa) of the trinomial $f(x)$ when x varies from negative infinity to positive infinity and passes by the two roots x_1 and x_2.

A **sign table** of the trinomial $f(x)$ can be set up showing the sign status of $f(x)$ within each interval of variation of x marked by the two roots. The sign table has many applications, one is to solve systems of quadratic inequalities.
Note. There is another approach, solving quadratic inequalities by graphing, which will be covered in Chapter 4.

3-2. SOLVING QUADRATIC INEQUALITIES BY THE NUMBER LINE.

Given the standard form of quadratic inequality, in which f(x) is a trinomial of variable x and a, b, c are real numbers with a not equal to zero,

$$f(x) = ax^2 + bx + c > 0 \quad \text{or} \quad f(x) = ax^2 + bx + c < 0 \qquad (1)$$

If the inequality is in another form, transform it into standard form by using common transformations such as factoring, moving and combining like terms.

Suppose x_1 and x_2 are the two real roots of equation $f(x) = 0$. Their graphs divide the number line into 3 parts, 2 rays and a segment. Any ray or segment whose coordinates make the inequality (1) true is part of the solution set.

Solving steps. To solve a quadratic inequality by a number line, use the following steps:

<u>Step 1.</u> Find the real roots x_1 and x_2 of the equation $f(x) = 0$. If the equation does not have real roots, the inequality either has no solution or is true for all values of x.

<u>Step 2.</u> Plot these 2 roots on the number line. These points divide the number line into 3 parts, one segment (x_1, x_2) and two rays $(-\infty, x_1)$, $(x_2, +\infty)$. The solution set is either the segment or the two rays.

<u>Step 3.</u> Replace the zero coordinate of the origin into the inequality (1). If it is true, then the origin is located on the solution set. If it is not true, then the origin is located outside the solution set.

<u>Example 1.</u> Solve the inequality $\quad f(x) = 4x^2 - 5x - 9 < 0 \quad (1)$

Solution. First solve $f(x) = 0$. Its two real roots are (-1) and $(2\frac{1}{4})$. Plot them as two open little circles on the number line. Replace $x = 0$ in inequality (1). It is true, so the origin is on the solution set which is the interval $(-1, 2\frac{1}{4})$

Important note. We recommend students to express the solution set with the interval ' notations . By convention we write the solution set in the form

a. $(-\infty, x_1)$ if the edge point x_1 is not included
b. $(-\infty, x_1]$ if x_1 is included
c. an open interval (x_1, x_2) if x_1, x_2 are not included
d. a closed interval $[x_1, x_2]$ if x_1, x_2 are included.

Example 2. Solve the inequality $\quad f(x) = 3x^2 - 4x - 7 \geq 0$

Solution. Solve $f(x) = 0$. Two real roots (-1) and $(2\frac{1}{3})$. Plot them on the number line. Replace $x = 0$ in inequality (1). It is untrue, so the origin is outside the solution set. The solution set is $(-\infty, -1]$ or $[2\frac{1}{3}, +\infty)$. The two roots are included.

Exercises. Solve and graph on a number line:

1. $5x^2 - 9x - 14 > 0$
2. $-x^2 + 6x + 16 < 0$
3. $3x^2 + 32x - 11 < 0$
4. $5x^2 - 11x + 2 > 0$
5. $-8x^2 + 22x - 15 < 0$
6. $6x^2 - x - 35 > 0$
7. $-3x^2 - 7x + 6 > 0$
8. $3x^2 - 7x + 9 < 0$
9. $15x^2 + x - 28 \geq 0$
10. $-12x^2 + 44x - 35 \leq 0$
11. $2x^2 - 6x + 3 \geq 0$
12. $3x^2 - 4x - 1 \geq 0$

3-3. SOLVING QUADRATIC INEQUALITIES BY THE RULE OF SIGN

3-3-1. Relationship between sign and roots of a trinomial.

The standard form of a quadratic inequality, in which $f(x)$ is a trinomial of variable x and a, b, c, are real numbers with $a \neq 0$, is :

$$f(x) = ax^2 + bx + c > 0 \quad \text{or} \quad f(x) = ax^2 + bx + c < 0 \quad (1)$$

Suppose D (or D') is the discriminant and x_1, x_2 are the two real roots of the equation $f(x) = 0$. There is a relationship between the discriminant and the sign of $f(x)$ in terms of the sign of a, as the variable x moves within its range and passes by the roots x_1 and x_2. The relationship is given by the following theorem.

3-3-2. Theorem on the sign status of a trinomial

Given a trinomial $f(x) = ax^2 + bx + c$ ($a \neq 0$) and the Discriminant $D = b^2 - 4ac$,

If $D < 0$ then $f(x)$ has the same sign as a for all values of x.

If $D = 0$ then $f(x)$ has the same sign as a for any value of $x \neq -\dfrac{b}{2a}$

If $D > 0$, $f(x)$ has two real roots x_1 and x_2, then the trinomial $f(x)$ has the same sign as a if x is outside the interval (x_1, x_2) and $f(x)$ has the opposite sign of a when x varies within the interval (x_1, x_2)

Note. We state the theorem without proving it. Students can find its proof in the Review Exercises and a graphic interpretation of the Theorem in Chapter 4, Paragraph 4-2.

3-3-3. The rule of sign and the sign table of a trinomial.

The rule of sign can be summarized as follows: between the real roots, $f(x)$ has the opposite sign of a. To illustrate the theorem, a sign table of $f(x)$ can be set up showing the sign status of $f(x)$ in three intervals as follows.

x	$-\infty$	x_1	x_2	$+\infty$
f(x)	Same sign as a	Opposite sign of a	Same sign as a	

51

3-3-4. Application of the rule of sign and the sign table

The rule of signs for trinomials is used to solve a single quadratic inequality. The sign table is used to solve systems of two or more quadratic inequalities. By considering the sign of f(x) in each interval, students can find the solution set easily as x varies from negative infinity to positive infinity.

<u>Example 1</u>. Solve the quadratic inequality $-5x^2 + 2x + 7 < 0$

Solution. The real roots are (-1) and ($1\frac{2}{5}$). Between the roots, f(x) is positive, the opposite sign of a.
The solution set is outside the interval marked by the two roots.
The solution set is (-∞, -1) or ($1\frac{2}{5}$, +∞) where f(x) is negative.

<u>Example 2</u>. Solve the inequality $3x^2 - 7x + 10 > 0$

Solution. Two real roots are (-1) and ($3\frac{1}{3}$). Between the roots, the trinomial is negative, the opposite sign of a
The solution set is (-∞, -1) or ($3\frac{1}{3}$, +∞) where f(x) is positive

<u>Example 3</u>. Solve the inequality $-x^2 - 5x + 6 \leq 0$

Solution. The two real roots are (1) and (-6) . Between the roots, the trinomial is positive, the opposite sign of a.
The solution set is (-∞, -6] or [1 , +∞). The two roots are included.

Exercises.   Using the rule of sign to solve:

1. $f(x) = 2x^2 + 9x + 7 > 0$ 2. $f(x) = x^2 + x - 6 < 0$

3. $f(x) = 3x^2 - 10x + 3 < 0$ 4. $f(x) = 3x^2 - 12x + 9 > 0$

5. $f(x) = -8x^2 + 22x - 15 < 0$ 6. $f(x) = -2x^2 - 13x - 11 > 0$

7. $f(x) = -6x^2 + 7x + 20 > 0$ 8. $f(x) = 2x^2 + 7x - 4 < 0$

9. $f(x) = 15x^2 + 22x + 8 \geq 0$ 10. $f(x) = 21x^2 - 23x + 6 \geq 0$

11. $f(x) = 2x^2 + 12x + 17 \leq 0$ 12. $f(x) = 2x^2 - 2x - 1 \leq 0$

3-4. SOLVING SYSTEMS OF QUADRATIC INEQUALITIES.

The standard form of a system of inequalities is

$$f(x) = ax^2 + bx + c \geq 0 \quad (a \neq 0) \quad (1)$$
$$g(x) = cx^2 + dx + e \geq 0 \quad (c \neq 0) \quad (2)$$

in which f(x) and g(x) are two trinomials of variable x and where a, b, c, d, e, f are real numbers. If f(x) and g(x) are in other forms, transform them to into standard form by using common transformations.
Solving this system of inequalities is to find any solution set of x that make both inequalities true. We explain here two common approaches to solve a system of inequalities.

    1. The first approach uses the number line and the real roots of the two equations f(x) = 0 and g(x) = 0.
    2. The second approach uses the real roots and the sign table of the two trinomials f(x) and g(x).

3-4-1. Solving systems of quadratic inequalities by using the number line.

Given a system of inequalities, where f(x) and g(x) are two trinomials of the variable x.

$$f(x) = ax^2 + bx + c \geq 0 \quad (a \neq 0) \quad (1)$$
$$g(x) = dx^2 + ex + f \geq 0 \quad (d \neq 0) \quad (2)$$

If f(x) and g(x) are in other forms, transform them into standard form.

To solve this system by a number line, use the following steps.

    <u>Step 1</u>. Find the real roots of two equations f(x) = 0 and g(x) = 0. Suppose x_1, x_2, x_3 and x_4 are the four real roots.
    <u>Step 2</u>. Plots the 4 roots on the number line. To avoid confusion and to easily see the solution set, we recommend the use of a double-line. Plot the roots of f(x) on the first line and the roots of g(x) on the second line. Both lines must have the same origin and distance marks. Use different colors for the two lines.
    <u>Step 3</u>. Find on each line the solution set by replacing the coordinate zero of the origin into inequalities (1) and (2). By superimposing, students can easily see the solution set that makes both inequalities true.

<u>Example 1.</u> Solve and graph this system on a number line

$$f(x) = 3x^2 - 7x + 4 > 0 \quad (1)$$
$$g(x) = 4x^2 - 5x - 9 < 0 \quad (2)$$

Solution. Solve $f(x) = 0$ Solve $g(x) = 0$
Tip case $a + b + c = 0$ Tip case $a - b + c = 0$

Two roots (1) and $(1\frac{1}{3})$ Two roots (-1) and $(2\frac{1}{4})$

Replace $x = 0$ in (1). It is true.   Replace $x = 0$ in (2). It is true.
The origin is out side the roots.   The origin is between the roots.

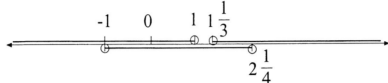

By superimposing, the solution set can be easily seen as $(-1, 1)$ or $(1\frac{1}{3}, 2\frac{1}{4})$

Example 2. Solve and graph $f(x) = 3x^2 - 4x - 7 > 0$ (1)
$g(x) = x^2 - x - 6 \geq 0$ (2)

Solution. Solve $f(x) = 0$ Solve $g(x) = 0$

Two roots (-1) and $(2\frac{1}{3})$ Two roots (-2) and (3)

The origin is outside the solution set   The origin is outside the solution set

The solution set of the system is $(-\infty, -2]$ and $[3, +\infty)$. The edge points are included.

Example 3. Solve and graph $f(x) = -3x^2 + 5x + 8 > 0$ (1)
$g(x) = x^2 - x + 9 > 0$ (2)

Solution. Solve $f(x) = 0$ Solve $g(x) = 0$

Case $a - b + c = 0$ $D < 0$. There are no real roots

Two roots (-1) $(2\frac{2}{3})$ $g(x)$ has same positive sign as a.

Replace $x = 0$ into (1). It is true   It is true for all values of x.

The origin is on the solution set which is the interval $(-1, 2\frac{2}{3})$

Exercises. Solve and graph on a number line:

1. $-2x^2 + 13x - 11 > 0$
    $3x^2 - 5x - 8 < 0$

2. $8x^2 - 2x - 15 > 0$
    $2x^2 + 6x - 8 > 0$

3. $x^2 - x - 6 < 0$
    $3x^2 - 6x - 9 > 0$

4. $10x^2 + 11x - 35 > 0$
    $x^2 + x - 12 < 0$

5. $2x^2 + 7x - 4 < 0$
    $6x^2 + 15x + 9 < 0$

6. $6x^2 - x - 35 < 0$
    $-2x^2 + 5x + 7 > 0$

7. $2x^2 + 7x - 4 > 0$
    $5x^2 + 9x - 2 < 0$

8. $8x^2 + 22x - 21 > 0$
    $3x^2 - 8x - 11 < 0$

9. $3x^2 + 2x - 21 \leq 0$
    $15x^2 - 4x - 35 \leq 0$

10. $16x^2 - 34x - 15 \leq 0$
    $3x^2 + 4x - 4 \geq 0$

3-4-2. Solving systems of inequalities using the sign table of trinomials

Given a system of inequalities, in which f(x) and g(x) are two trinomials of variable x, and a, b, c, d, e, f, are real numbers,

$$f(x) = ax^2 + bx + c \geq 0 \quad (a \neq 0)$$
$$g(x) = dx^2 + ex + f \geq 0 \quad (d \neq 0)$$

Transform them into standard form if they are in another form.
Use the following steps to solve:
  Step 1. Find the 4 real roots of the two equations f(x) = 0 and g(x) = 0. If one equation does not have real roots, use the Theorem to find its sign status.
  Step 2. Set up the sign table of f(x) and g(x)
  Step 3. Find the intervals where both inequalities are true.

Example 1. Solve system
$$f(x) = 3x^2 - 5x + 2 < 0 \quad (1)$$
$$g(x) = x^2 - 7x + 6 > 0 \quad (2)$$

Solution. Solve f(x) = 0
Two real roots (1) and ($\frac{2}{3}$)

Solve g(x) = 0
Two real roots (1) and (6)

Setup the sign table.

x	$-\infty$		2/3		1		6		$+\infty$
f(x)		+	0	-	0	+		+	
g(x)		+		+	0	-	0	+	
System		no		yes		no		no	

The solution set is easily seen. It is the interval $(\frac{2}{3}, 1)$ where f(x) is negative and g(x) is positive. Solving by using the number line, we get the same result.

Example 2. Solve system $\quad$ f(x) = $-x^2 - 5x + 6 < 0$
$\qquad\qquad\qquad\qquad\qquad$ g(x) = $3x^2 - 7x - 10 > 0$

Solution. Solve f(x) = 0 $\qquad\qquad\qquad$ Solve g(x) = 0

Two roots (1) and (-6) $\qquad\qquad\qquad$ Two roots (-1) and $(3\frac{1}{3})$
Set up the sign table

x	$-\infty$		-6		-1		1		$3\frac{1}{3}$		$+\infty$
f(x)		-	0	+		+	0	-		-	
g(x)		+		+	0	-		-	0	+	
System		yes		no		no		no		yes	

The solution set are $(-\infty, -6)$, $(3\frac{1}{3}, +\infty)$.

Example 3. Solve the system $\quad 7x - 15 \geq -2x(x+3)$ $\qquad$ (1)
$\qquad\qquad\qquad\qquad\qquad\qquad 3x(x-3) \leq 12$ $\qquad\qquad\quad$ (2)

Solution. Transform inequalities (1) and (2) into standard form:

$$f(x) = 2x^2 + 13x - 15 \geq 0$$
$$g(x) = 3x^2 - 9x - 12 \leq 0$$

Solve f(x) = 0   Solve g(x) = 0
Two real roots (1) $(-7\frac{1}{2})$ Two real roots (-1) (4)

Set up the sign table

x	-∞		$-7\frac{1}{2}$		-1		1		4		+∞
f(x)		+	0	-		-	0	+		+	
g(x)		+		+	0	-		-	0	+	
Syst.		no		no		no		yes		no	

The solution set is [1, 4] where f(x) is positive and g(x) is negative. The edge points are included.

Exercises.   Solve these systems by using a sign table:

1. $7x^2 - 10x - 17 > 0$
   $-x^2 + 2x + 24 < 0$

2. $15x^2 + x - 28 > 0$
   $20x^2 - 31x + 12 > 0$

3. $-12x^2 + 43x - 35 > 0$
   $35x^2 + 24x - 27 < 0$

4. $21x^2 + 26x - 15 < 0$
   $15x^2 - 4x - 35 > 0$

5. $x^2 - 6x - 7 \leq 0$
   $x^2 + x - 6 \leq 0$

6. $5x^2 - 8x + 3 \leq 0$
   $x^2 + 2x - 8 \geq 0$

7. $x^2 + 4x - 21 > 0$
   $-4x^2 + 39x - 56 < 0$

8. $6x^2 + x - 12 < 0$
   $15x^2 + x - 40 < 0$

9. $6x^2 + 29x + 35 > 0$
   $x^2 - 9x + 14 < 0$

10. $35x^2 - 11x - 6 < 0$
    $-21x^2 + 13x + 20 < 0$

11. $6x^2 + 13x + 5 \leq 0$
    $-4x^2 + 39x - 56 \leq 0$

12. $8x^2 + 22x - 21 \geq 0$
    $-6x^2 + 19x + 77 \geq 0$

3-5. SOLVING OTHER SYSTEMS OF QUADRATIC INEQUALITIES

The method of using a sign table can be used to solve other types of inequality systems:
. Mixed systems of linear and quadratic inequalities
. System of three or more quadratic inequalities
. Systems with exponents in x higher than two that can be transformed into quadratic inequalities.

3-5-1. Mixed system of linear and quadratic inequalities.

The Sign Table method, solving by intervals, is very convenient in solving systems that are mixed systems of linear and quadratic inequalities

Example 1. Solve the system $\quad f(x) = 2x - 3 < 0$
$\quad\quad\quad\quad\quad\quad\quad\quad\quad\quad\quad\quad g(x) = x^2 + x - 6 > 0$

Solution.

Solve $f(x) = 0$ $\quad\quad\quad\quad\quad\quad\quad\quad$ Solve $g(x) = 0$
One root ($\frac{3}{2}$) $\quad\quad\quad\quad\quad\quad\quad\quad$ Two roots (-3) $(+2)$

Set up the sign table.

x	$-\infty$		-3		$\frac{3}{2}$		$+2$		$+\infty$
f(x)		$-$		$-$	0	$+$		$+$	
g(x)		$+$	0	$-$		$-$	0	$+$	
System		yes		no		no		no	

The solution set is $(-\infty, -3)$ where $f(x)$ is negative and $g(x)$ is positive.

Example 2. Solve system $f(x) = 3 - 2x > 0$ (1)
$g(x) = x^2 - 6x + 7 < 0$ (2)
$h(x) = -x^2 + 3x + 4 > 0$ (3)

Solution. Solve $f(x) = 0$.   Root: $(\frac{3}{2})$
              $g(x) = 0$ Two real roots (1) (7)

              $h(x) = 0$ Two real roots (-1) (4).

Set up the sign table

x	$-\infty$		-1		1		$\frac{3}{2}$		4		7		$+\infty$
f(x)		+		+		+	0	-		-		-		
g(x)		+		+	0	-		-		-	0	+		
h(x)		-	0	+		+		+	0	-		-		
Syst.		no		no		yes		no		no		no		

The solution set is $(1, \frac{3}{2})$. In this interval, g(x) is negative, f(x) and h(x) are both positive.

3-5-2. System of three or more quadratic inequalities

The method of using a sign table is convenient in solving systems of three or more quadratic inequalities

Example 1. Solve the system of three quadratic inequalities:

$2x^2 + 5x - 8 < x^2 + 3x - 5$ transforms into $f(x) = x^2 + 2x - 3 < 0$
$-x^2 - 2x + 5 > 2x^2 - 9x + 7$ transforms into $g(x) = -3x^2 + 7x - 2 > 0$
$-3x^2 + 2x + 3 < -5x^2 + 11x + 14$ transforms into $h(x) = 2x^2 - 9x - 11 < 0$

Solution. Solve the equations $f(x) = 0$, $g(x) = 0$, and $h(x) = 0$.

Roots are respectively,   (1), (-3); $(\frac{1}{3})$, (2);   (-1), $(5\frac{1}{2})$

Set up the sign table

x	$-\infty$		-3		-1		$\frac{1}{3}$		1		2		$5\frac{1}{2}$		$+\infty$
f(x)		+	0	-		-		-	0	+		+		+	
g(x)		-		-		-	0	+		+	0	-		-	
h(x)		+		+	0	-		-		-		-	0	+	
Sys		no		no		no		yes		no		no		no	

The solution set is ($\frac{1}{3}$, 1) where g(x) is positive and f(x) & h(x) are both negative.

Exercises. Set up the sign table and solve

1. System
$- x^2 + 2x + 3 < 2x (x - 3) + 7$
$4x^2 - x + 2 > - 3x^2 + 2x - 2$
$- x^2 + x + 8 < - 3x^2 + 8x + 17$

2. System
$5x^2 - 3x + 3 < x^2 - 11x + 15$
$- x^2 + 3x - 5 < - 2x^2 + 5x + 3$
$- 2x^2 + 5x + 3 > x^2 + x - 4$

3. System
$f(x) = 15x^2 + 10x - 40 < 0$
$g(x) = - 6x^2 + 7x + 20 > 0$
$h(x) = 20x^2 + 7x - 3 < 0$

4. System
$f(x) = 15x^2 + 94x - 40 \leq 0$
$g(x) = 20x^2 - 67x + 56 \geq 0$
$h(x) = - 6x^2 + 19x + 77 \geq 0$

3-5-3. System of inequalities that can be transformed into quadratic inequalities

The Sign Table Method can be used to solve many inequalities with degrees of x higher than two if these inequalities can be transformed into linear and/or quadratic inequalities. First, it is useful to know the following Tips.

TIPS. For solving a cubic equation $ax^3 + bx^2 + cx + d = 0$ quickly, students are advised to remember these following tips.

Tip 1. If $a + b + c + d = 0$, one real root is 1
For example, the equation $3x^3 - 5x^2 + 7x - 5 = 0$ has one real root equal to 1

Tip 2. If $a - b + c - d = 0$, one real root is -1
For example, the equation $4x^3 + 8x^2 + 9x + 5 = 0$ has one real root equal to -1

<u>Example 1</u>. Solve the system $f(x) = x^3 - 7x^2 + 14x - 8 > 0$ (1)
$g(x) = -3x^2 - 7x + 6 < 0$ (2)

Solution. The equation $f(x)$ has 1 as a real root by application of Tip 1.

Factor out $(x - 1)$ $f(x) = (x - 1)(x^2 - 6x + 8) > 0$. Let $h(x) = x^2 - 6x + 8$

Solve $f(x) = 0$ Three real roots   (1), (2), (4)

Solve $g(x) = 0$ Two real roots (-3), $(\frac{2}{3})$
Set up the sign table:

x	$-\infty$	-3		$\frac{2}{3}$		1		2		4		$+\infty$
x - 1	-		-		-	0	+		+		+	
h(x)	+		+		+		+	0	-	0	+	
f(x)	-		-		-	0	+	0	-	0	+	
g(x)	-	0	+	0	-		-		-		-	
Syst.	no		no		no		yes		no		yes	

The solution set is $(1, 2), (4, +\infty)$ where $f(x)$ is positive and $g(x)$ is negative.

<u>Example 2</u>. Solve system $f(x) = x^4 - 3x^2 - 4 < 0$
$g(x) = x^2 + 2x - 3 > 0$

Solution. Let $x^2 = X$, $f(x)$ becomes a quadratic equation which can be factored

$f(x) = X^2 - 3X - 4 = (X + 1)(X - 4) = (x^2 + 1)(x^2 - 4)$

Solve   f(x) = 0   Roots (-2) and (2)
         g(x) = 0   Roots (-3) and (1)

Set up the sign table

x	-∞	-3	-2	1	2	+∞
x² + 1	+	+	+	+	+	
x² - 4	+	+	0 -	-	0 +	
f(x)	+	+	0 -	-	0 +	
g(x)	+	0 -	-	0 +	+	
Syst.	No	No	No	yes	No	

The solution set is (1, 2) where f(x) is negative and g(x) is positive.

Exercises.   Solve the system

1. f(x) = x³ - 9x² + 20x – 12 > 0
    g(x) = x² - x – 6 < 0
    Hint: Tip 1

2. f(x) = x³ - 21x + 20 > 0
    g(x) = x² - 8x + 12 < 0
    Hint: Tip 1

3. f(x) = x³ - 8x² + 5x + 14 > 0
    g(x) = 2x² + 7x - 4 < 0
    (Hint: Tip 2)

4. f(x) = 5x³ + 14x² + 7x – 2 < 0
    g(x) = -21x² + 13x + 20 > 0
    (Hint: Tip 2)

5. f(x) = 2x³ + 5x² - 11x + 4 > 0
    g(x) = 6x² + 13x + 5 < 0
    (Hint: Tip 1)

6. f(x) = (x² - 4)(3 – x) > 0
    g(x) = 6x² - x – 35 < 0

7. f(x) = x⁴ - 4x² + 5 < 0
    g(x) = 12x² - 17x – 7 < 0

8. f(x) = x⁴ - 8x² - 9 < 0
    g(x) = 5x² + 9x – 2 < 0

Given   f(x) = 5x² + 4x - 9   and   g(x) = 12x² - 17x - 7

9. Solve $\dfrac{f(x)}{g(x)} > 0$

10. Solve $\dfrac{f(x)}{g(x)} < 0$

3-6. CHAPTER REVIEW. Solving quadratic inequalities

The standard form of a linear inequality, in which f(x) is a trinomial of variable x and a, b, c are real numbers with a not equal to zero, is:

$$f(x) = ax^2 + bx + c > 0 \quad \text{or} \quad f(x) = ax^2 + bx + c < 0 \quad (1)$$

If the inequality is in another form, transform it into standard form by common transformations.

There are two common approaches to solve a quadratic inequality

The first approach uses the number line and the two real roots of the equation f(x) = 0. The zero coordinate of the origin is replaced into inequality (1) to see if it makes the inequality true. By consequence, the solution set can be found.

The second approach uses the two real roots of the equation f(x) = 0 and the rule of sign. The rule of sign is based on a theorem which states that if x is located between the 2 roots, then f(x) has the opposite sign of the coefficient a. The sign table of f(x) shows its sign status in each interval when x varies from negative infinity to positive infinity and passes by the two real roots.

Solving systems of quadratic inequalities

The standard form of a system of quadratic inequalities is in which f(x) and g(x) are two trinomials of variable x, and a, b, c, d, e, f are real numbers,

$$\begin{array}{lll} f(x) = ax^2 + bx + c \geq 0 & (a \neq 0) & (1) \\ g(x) = dx^2 + ex + f \geq 0 & (d \neq 0) & (2) \end{array}$$

The system should be transformed to standard form if it is in another form.

There are two common approaches to solve a system of quadratic inequalities.

a. The first approach uses the number line and the four real roots of the two equations f(x) = 0 and g(x) = 0. These four coordinates are plotted on a double number line.

The coordinate of the origin is replaced respectively into the two inequalities (1) and (2) to find two separate solution sets. By superimposing the two results on the double number line, the solution set of the system can be easily found.

b. The second approach uses the four roots and the sign table of f(x) and g(x) as x varies from negative infinity to positive infinity and passes by the four roots. The sign status of f(x) and g(x) are obtained by applying the theorem on the sign status of a trinomial.

The method of using a sign table can be used to solve systems of multiple-quadratic inequalities. It is also used to solve systems with degrees of x higher than 2 if these systems can be transformed into quadratic inequalities.

3-7. REVIEW EXERSICES

a. Solving an inequality by the number line

1. $-6x^2 + 19x + 77 > 0$
2. $15x^2 - 56x + 49 < 0$
3. $5x^2 - 28x + 15 > 0$
4. $2x^2 - 9x + 9 > 0$
5. $-4x^2 + 39x - 56 > 0$
6. $15x^2 + x - 40 < 0$
7. $-3x^2 + 5x + 12 < 0$
8. $8x^2 + 26x + 15 < 0$
9. $16x^2 - 34x - 15 \geq 0$
10. $-4x^2 + 39x - 56 \leq 0$

b. Solving a quadratic inequality by the rule of sign

1. $12x^2 - 39x + 30 > 0$
2. $5x^2 - 8x - 21 < 0$
3. $7x^2 - 41x + 30 < 0$
4. $21x^2 - 23x - 20 > 0$
5. $-5x^2 + 37x - 42 > 0$
6. $11x^2 + 32x - 3 < 0$
7. $15x^2 - 51x + 42 > 0$
8. $21x^2 - 23x + 6 < 0$
9. $6x^2 + 13x + 5 \leq 0$
10. $-7x^2 + 25x - 12 \geq 0$
11. $3x^2 + 4x - 1 \leq 0$
12. $2x^2 + 6x + 2 \geq 0$

c. Solving a system of quadratic inequalities by the number line

1. $3x^2 - 4x - 7 < 0$
   $-12x^2 + 43x - 35 > 0$
2. $2x^2 - 9x + 9 < 0$
   $16x^2 - 34x - 15 < 0$
3. $15x^2 + x - 40 < 0$
   $3x^2 - 7x - 10 > 0$
4. $x^2 - 7x + 6 > 0$
   $-6x^2 + 19x + 77 > 0$
5. $6x^2 + 13x + 5 > 0$
   $21x^2 - 48x - 45 < 0$
6. $-3x^2 + 7x - 2 > 0$
   $15x^2 + x - 40 > 0$
7. $8x^2 + 22x - 21 < 0$
   $2x^2 + 7x - 9 > 0$
8. $6x^2 + 13x + 5 > 0$
   $8x^2 + 7x - 1 < 0$
9. $-4x^2 + 39x - 56 \geq 0$
   $x^2 - 4x + 1 \geq 0$
10. $x^2 - 9x + 14 \leq 0$
    $2x^2 - 8x + 7 \leq 0$

d. Solving a system of quadratic inequalities by the sign table

1. $2x^2 - 8x + 5 > 0$
    $8x^2 - 2x - 15 < 0$

2. $x^2 - 4x + 2 > 0$
    $2x^2 - 9x + 9 > 0$

3. $4x^2 - 9x - 9 < 0$
    $3x^2 - 20x + 12 < 0$

4. $-3x^2 - x + 14 < 0$
    $x^2 - 7x - 8 < 0$

5. $5x^2 - 2x - 16 > 0$
    $-7x^2 + 26x - 15 > 0$

6. $14x^2 - 15x - 9 > 0$
    $-x^2 + 7x + 8 < 0$

7. $x^2 - 5x + 5 \geq 0$
    $-7x^2 + 78x - 11 \geq 0$

8. $-3x^2 + 22x - 7 \leq 0$
    $7x^2 + 9x + 2 \leq 0$

e. Solving a system of multiple quadratic inequalities

1. $9x^2 + 3x - 20 < 0$
    $7x^2 - 78x + 11 > 0$
    $3x^2 - 4x - 7 < 0$

2. $11x^2 - 29x - 12 < 0$
    $8x^2 - 24x + 10 > 0$
    $12x^2 + 29x + 15 > 0$

3. $10x^2 - 7 > 2x^2 + 26x - 22$
    $-5x^2 + 3x - 2 < 20x - 14$
    $4x^2 + 3x - 2 > x^2 + 8x + 6$

4. $2x^2 - 2x + 13 > 3x^2 - 5x + 3$
    $4x^2 + 2 > x^2 + 13x - 10$
    $11x^2 + 7x + 2 < 10x + 16$

f. Solving a system of inequalities which can be transformed into quadratic inequalities.

1. $x^3 - 8x^2 + 17x - 10 < 0$
    $x^2 - 5x - 6 < 0$

2. $-x^3 + 7x^2 - 7x - 15 < 0$
    $3x^2 - 16x + 5 > 0$

3. $3x^3 - 8x^2 - 3x + 8 < 0$
    $x^2 - x - 6 < 0$

4. $(x^2 - 8x - 9)(3x^2 - 16x + 5) < 0$
    $x^2 - 6x - 7 > 0$

g. Exercises on proofs of the rule of sign for a trinomial.

1. Given a trinomial $f(x) = ax^2 + bx + c$ with a not equal to zero. Prove that if the discriminant D is negative (D < 0), then f(x) has the same sign as a for all values of x.

Hint: Write f(x) in the form $f(x) = a[(x + \frac{b}{2a})^2 - \frac{D}{4a^2}]$

2. Prove that if D = 0, then f(x) has the same sign as a for all values of x except when $x = -\frac{b}{2a}$

Hint: Write f(x) in the form $f(x) = a(x + \frac{b}{2a})^2$

3. Prove that if D > 0, then f(x) has the opposite sign of a when x varies within the interval marked by the two real roots x_1 and x_2 of the equation f(x) = 0.

Hint: Write f(x) in the form $f(x) = a(x - x_1)(x - x_2)$. Then set up a sign table for the two binomials $(x - x_1)$ and $(x - x_2)$ and their product.

CHAPTER 4

SOLVING QUADRATIC INEQUALITIES BY GRAPHING

4-1. THE QUADRATIC FUNCTION $y = f(x) = ax^2 + bx + c$

In order to learn how to solve a quadratic inequality by graphing, let's review the features of a quadratic function and its graph.

Study of a quadratic function.

The standard form (general form) of a quadratic function is: $y = ax^2 + bx + c$ in which x is the independent variable and a, b, c are constants with $a \neq 0$

a. Definition.
The function is defined for all real values of x when it varies in its domain from negative infinity to positive infinity.

b. Graph.
The graph of a quadratic function is called a **parabola**.

If **a > 0**, the parabola is **upward**, meaning y goes to positive infinity (+∞) when x goes to negative or positive infinity.

For example, the graph of the function $y = f(x) = 3x^2 + 4x - 5$ and the graph of the function $y = g(x) = 4x^2 + 2x - 1$ are upward parabolas.

If **a < 0**, the parabola is **downward**, meaning y goes to negative infinity (-∞) when x goes to negative or positive infinity.

For example, the graphs of the functions $y = f(x) = -x^2 + 4x - 7$ and $y = g(x) = -2x^2 + 3x - 5$ are downward parabolas.

c. Axis of symmetry.

The parabola has an axis of symmetry which is the line parallel to the y-axis at

$$x = -\frac{b}{2a}$$

Example 1. The function $y = 3x^2 - 6x + 1$ has an axis of symmetry which is parallel to the y-axis at $x = -\frac{(-6)}{2(3)} = 1$

Example 2. The function $y = f(x) = -3x^2 + 7x - 2$ has an axis of symmetry which is parallel to the y-axis at $x = \frac{-7}{2(-3)} = \frac{7}{6}$

Remark. If $c = 0$, the function is $y = x(ax + b)$. The graph is a parabola intersecting the x-axis at the origin and at point $-\frac{b}{a}$. The axis of symmetry is the line parallel to the y-axis at $x = -\frac{b}{2a}$

d. Maximum or Minimum.

When $x = -\frac{b}{2a}$, we have $y = -\frac{D}{4a}$ with $D = b^2 - 4ac$

If $a > 0$, the vertex $(-\frac{b}{2a}, -\frac{D}{4a})$ is a **minimum,** meaning the vertex is the lowest point on the graph

If $a < 0$, the vertex is a **maximum** meaning the vertex is the highest point on the graph.

Example 1. The function $y = x^2 + 3x - 4$ has a minimum at $x = -\frac{b}{2a} = -\frac{-3}{2} = \frac{3}{2}$

and it equals to $y = -\frac{D}{4a} = -\frac{25}{4} = -6\frac{1}{4}$

Example 2. The function $y = -x^2 + 2x + 3$ has a maximum $y = -\frac{D}{4a} = 4$

at $x = \frac{-2}{-2} = 1$

e. Axis-intercepts.

To find the **y-intercept**, replace $x = 0$ in the function's equation. We get $y = c$. By symmetry, we can get another corresponding point on the graph.

To find the **x-intercepts**, make $y = 0$. The x-intercepts are the real roots of the equation

$$y = ax^2 + bx + c = 0 \qquad (1)$$

. If the Discriminant $D = b^2 - 4ac < 0$, the graph does not intersect the x-axis

. If $D = 0$, the graph intersects the x-axis at one point at $x = -\dfrac{b}{2a}$

. If $D > 0$, the graph intersects the x-axis at two points whose x-coordinates are the real roots of the equation (1).

Example 1. Find the axis-intercepts of the equation $f(x) = 4x^2 - 5x + 1$

Solution.
y-intercept   Make $x = 0$, we get $y = 1$
x intercepts: Make $y = 0$, and solve the equation $4x^2 - 5x + 1 = 0$ whose two real roots are (1) and $(\dfrac{1}{4})$

Example 2. Find the axis-intercepts of $f(x) = 5x^2 + 7x + 2$

Solution.

y-intercept: Make $x = 0$, we get $y = 2$

x-intercept: Make $y = 0$, and solve the equation $5x^2 + 7x + 2 = 0$ whose two real roots are (-1) and $(-\dfrac{2}{5})$

Note. If x_2 and x_1 are the two real roots of the equation (1), the x-coordinate of the vertex, or axis of symmetry, is the average $\dfrac{x_1 + x_2}{2} = -\dfrac{b}{2a}$

f. Table of variation.

Below are the two general forms of the variation table of any quadratic function. In order to graph the function, some more characteristic points or ordered pairs need to be added to the table.

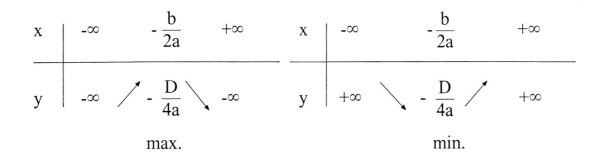

Example 1. Study and graph the function $y = -x^2 + 1$

Solution. The function is defined for all real values of x

Axis of symmetry: It is the y-axis since b = 0

y-intercept: $x = 0 \longleftrightarrow y = 1$

x-intercept: $y = 0 \longleftrightarrow x^2 = 1$ or $x = \pm 1$

Maximum vertex: Point (0, 1) on the y-axis
The function is increasing when x is in the interval $(-\infty, 0)$
It is decreasing in the interval $(0, +\infty)$

Variation table:

x	$-\infty$		-2		-1		0		1		2		$+\infty$
y	$-\infty$	↗	-3	↗	0	↗	1	↘	0	↘	-3	↘	$-\infty$
							max.						

The graph is shown on Figure 1.

Example 2.   Study and graph the function $y = 2x^2 - 4x + 2$

Solution. The function is defined for all real numbers of x

70

Axis of symmetry: The line $x = -\dfrac{(-4)}{2(2)} = 1$

Minimum vertex   Point $(1, 0)$ since when $x = 1 \longrightarrow y = 0$
The function is decreasing when x varies in the interval $(-\infty, 1)$
It is increasing when x varies in the interval $(1, +\infty)$

y-intercept: $x = 0 \longleftrightarrow y = 2$

x-intercept: $y = 0 \longleftrightarrow 2x^2 - 4x + 2 = 0$
                     One double root $x = 1$. One intersection at $x = 1$

Variation table

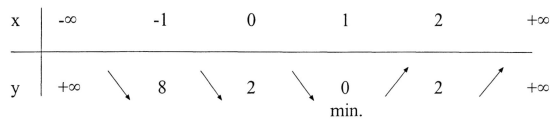

The graph is shown on Figure 2.

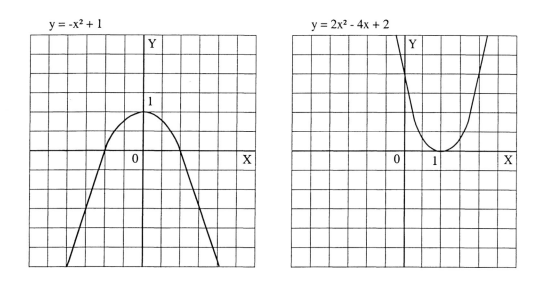

Figure 1.   Figure 2.

Exercises on graphing quadratic functions. Set up the variation table and graph the function

1. $y = 2x^2 - 1$ 2. $y = x^2 - 2$

3. $y = \dfrac{x^2}{2} - 1$ 4. $y = -2x^2 + 4$

5. $y = 2x^2 + 5x + 3$ 6. $y = x^2 - 4x + 3$

7. $y = x^2 - 3x - 4$ 8. $y = \dfrac{x^2}{2} + x + \dfrac{3}{2}$

9. $y = -x^2 + 3x - 2$ 10. $y = -x^2 + 4x - 3$

4-2. SOLVING A QUADRATIC INEQUALITY BY GRAPHING.

We have learned in Paragraph 3-3 on how to use the rule of sign to solve a quadratic inequality which is in standard form

$$f(x) = ax^2 + bx + c \gtrless 0 \quad (1)$$

In this paragraph we will see how the inequality (1) is related to the graph of the function f(x) and what can be seen as a graphic interpretation of the rule of sign for a trinomial.

Suppose **a > 0**, then the graph of the quadratic function f(x) is an upward parabola. See Figure 3 .

This parabola intersects the x-axis at two points A and B whose x-coordinates x_1, x_2 are the two real roots of the equation f(x) = 0.

Consider a point C on the graph between A and B. This point C is located **below** the x-axis and its y-coordinate is **negative**. All other similar points on this portion of the graph between A and B will have negative y-coordinates making the inequality (1) true.

We see that the rule of sign (Paragraph 3-3-2) is justified. When x varies in the interval between the two roots x_1 and x_2, the trinomial f(x) is negative having the opposite sign of a. See Figure 3.

Suppose **a < 0**, and the graph of the function f(x) is a downward parabola. See Figure 4. Consider a point C which is located on the graph **above** the x-axis and its y-coordinate is **positive**. All similar points on this portion of the graph between A and B will have positive y-coordinates making the inequality (1) true.

The rule of sign of a trinomial is again justified. Between the two roots, the trinomial f(x) is positive having the opposite sign of a. See Figure 4.

Figure 3   Figure 4

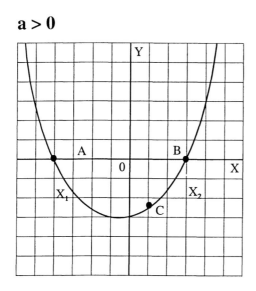
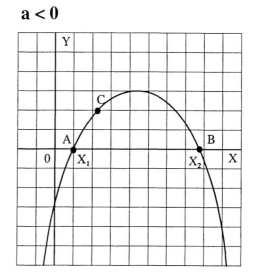

From the above studies, we can set up the steps to solve a quadratic inequality by graphing.

$$y = f(x) = ax^2 + bx + c \geq 0 \quad (1)$$

<u>Step 1.</u> Graph the function $y = f(x) = ax^2 + bx + c$

<u>Step 2.</u> Find the portion(s) of the graph, **above** or **below** the x-axis, so that the inequality (1) is true.

Note. With the help of graphing calculators, this method is convenient to solve systems of inequalities.

Example 1. Solve the inequality $f(x) = -x^2 + 1 < 0$

Solution. The graph of the function is shown on Figure 1
    On this figure, the graph of f(x) is **below** the x-axis, meaning f(x) is **negative**, when x is out side the interval marked by the two roots, (-1 , 1). It is justified by the rule of sign, f(x) has the same sign as a when x is out side the interval marked by the two roots.
As a consequence, the solution set is (-∞, -1)   or   (1 , +∞)

Example 2. Solve the inequality $f(x) = 2x^2 - 4x + 2 > 0$

Solution. The graph of this function is shown on Figure 2.

The graph of f(x) is completely above the x-axis so f(x) is **positive** for all values of x except when $x = -\frac{b}{2a} = 1$. The rule of sign in the case D = 0 is justified.

Exercises on solving a quadratic inequality by graphing.

Graph the functions then solve these inequalities. Compare the results with those from the exercises in Paragraph 3-3-4.

1. $f(x) = 2x^2 + 9x + 7 > 0$ 2. $f(x) = x^2 + x - 6 < 0$

3. $f(x) = 3x^2 - 10x + 3 < 0$ 4. $f(x) = 3x^2 - 12x + 9 > 0$

5. $f(x) = -8x^2 + 22x - 15 \leq 0$ 6. $f(x) = -2x^2 - 13x - 11 \geq 0$

7. $f(x) = -6x^2 + 7x + 20 > 0$ 8. $f(x) = 2x^2 + 7x - 40 < 0$

9. $f(x) = 15x^2 + 22x + 8 \geq 0$ 10. $f(x) = 2x^2 - 2x - 1 \geq 0$

4-3. SOLVING A SYSTEM OF QUADRATIC INEQUALITIES BY GRAPHING.

In Paragraph 3-4, we have learned about using a sign table to solve a system of quadratic inequalities that is in standard form

$$f(x) = ax^2 + bx + c \geq 0 \quad (1) \quad (a \neq 0)$$
$$g(x) = dx^2 + ex + f \geq 0 \quad (2) \quad (d \neq 0)$$

in which a, b, c, d, e, f, are real number constants. We are going to see the connection between the solution set with the graphs of the two functions f(x) and g(x). This connection can be considered as a graphic interpretation of the sign table.

Example 1. Solve the system $f(x) = x^2 + 3x - 4 > 0$
$g(x) = -x^2 + 2x + 3 > 0$

Solution. By using the sign table method (Paragraph 3-4) we can get the solution set of this system. It is the interval (1 , 3) where both f(x) and g(x) are positive.
Set up the variation table of the two functions
Variation table of $f(x) = x^2 + 3x - 4$

x	$-\infty$		-4		$-1\frac{1}{2}$		0		1		$+\infty$
y.	$+\infty$	↘	0	↘	$-6\frac{1}{4}$	↗	-4	↗	0	↗	$+\infty$

Variation table of $g(x) = -x^2 + 2x + 3$

x	$-\infty$		-1		0		1		3		$+\infty$
y	$+\infty$	↘	0	↘	3	↘	4	↗	0	↗	$+\infty$

The graphs of f(x) and g(x) are shown on Figure 5. We see that when x varies in the interval (1 , 3), both graphs are **above** the x-axis, meaning both functions f(x) and g(x) are **positive**. So, the interval (1 , 3) is the solution set of the system.

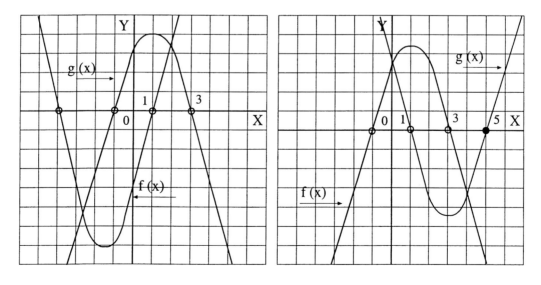

Figure 5   Figure 6

Example 2. Solve the system $f(x) = -x^2 + 2x + 3 < 0$
$g(x) = x^2 - 6x + 5 \leq 0$

Solution. First graph the two function f(x) and g(x).
The x-intercepts of the two graphs are real roots of the two equations
$f(x) = 0$ and $g(x) = 0$. They are respectively (-1), (3) ; (1) (5)
We see on Figure 6 that when x is within the interval (3 , 5), both graphs are **below** the x-axis meaning both functions are **negative.** By consequence, the solution set of the system is the interval (3 , 5). The x-intercept point, at x = 5, is included in the solution set.

Example 3 : Application to problem solving by graphing. The volume q in thousands of cubic meters of water pumped by a power station at a dam can be represented by the relation $q_1 = -t^2 + 6t$ in which t is the time given in hours. Another power station generates a volume which can be represented by the relation
$q_2 = -t^2 + 9t - 8$. In a time frame of 12 hours,

a. What is the period of time when both stations are working ?
b. What are the periods of time when only one station is working ?
c. What is the total water volume generated by both stations at time t = 3?

Solution. Study the variation and graph the two functions $f(t) = -t^2 + 6t$ and $g(t) = -t^2 + 9t - 8$ on the same coordinate plane. See Figure 7.

Variation table of $f(t) = -t^2 + 6t$

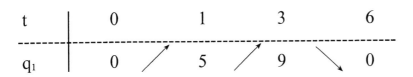

Variation table of $g(t) = -t^2 + 9t - 8$

t	0	1	$4\frac{1}{2}$	8	9
q_2	-8 ↗ 0 ↗ 12.25 ↘ 0 ↘ -8				

a. We see that within the interval (1, 6) both stations are working.
b. Within the intervals (0 , 1) and (6 , 8) there is only one station working.
c. The total generated volume is approximately 19 K m³ at time t = 3.

Figure 7   Figure 8

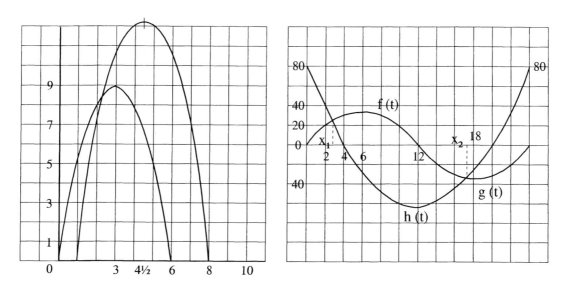

Example 4. Application to problem solving by graphing.

The tidal water of a swam pond rises up and down and run back and forth to and from the nearby ocean during a 24-hours period. The elevation EL1, in cm, of the pond water can be represented by the function $EL1 = f(t) = -t^2 + 12t$ from midnight to noon, and by the function $EL2 = g(t) = t^2 - 12t$ from noon to midnight in which t is the time given in hours. The elevation of the tidal ocean water can be represented by the function $ELO = h(t) = (t-4)(t-20)$. Water flows from high elevation to lower elevation. In a 24-hour time frame,
   a. What is the period of time when the water flows from the pond out to the ocean ?
   b. Estimate the one hour period when there is maximum volume of outflow ?
   c. Exactly at what moments does the water stop flowing ?
   d. What are the periods of time when there is water flowing from the ocean to the pond ?

Solution. To solve the problem by graphing, first graph the three functions on the same coordinate plane. See Figure 8.
The function f(t) intersects h(t) at x_1 . The function h(t) intersects g(t) at x_2 .
   a. Within the interval (x_1 , x_2) the pond water's elevation is higher than the one in the ocean, so there is water flowing from pond to ocean.
   b. There is maximum of water outflow when the difference of the two elevations is the highest, estimated at 80 cm, about between 9:00 to 10:00 am.

c. Exactly at the time x_1 and x_2, the water stops flowing because there is no difference in elevation.

We can get the exact value of x_1 by solving the equation:

$(t-4)(t-20) = -t^2 + 12t$	$f(t) = h(t)$
$2t^2 - 36t + 80 = 0$	Transform and simplify
At $t = 2.60$ or at 2: 36 am	Solve for t. Answer

We can get the exact value of x_2 by solving the equation:

$T^2 - 12T = (T-8)(T+8)$	Change to new y-axis with $T = t - 12$
$12T = 64$	Transform and simplify
$T = 5.33$	Solve for T
$t = 12 + 5.33 = 17.33$	Solve for t
At $t = 17: 20$ pm	Answer.

d. There is water running in reverse direction within the interval (0 , 2:36 am) and (17:20 pm , 24 pm) because the elevation of the ocean water is higher.

Exercises on solving systems of quadratic inequalities by graphing.

Graph the functions and solve these systems of inequalities.

1. $-x^2 + 3x + 4 > 0$
   $x^2 + 4x - 5 > 0$

2. $-x^2 - 5x + 6 < 0$
   $3x^2 - 7x - 10 < 0$

3. $2x^2 + 13x - 15 > 0$
   $3x^2 - 9x - 12 < 0$

4. $3x^2 - 4x - 7 > 0$
   $x^2 - x - 6 > 0$

5. $x^2 - 3x - 4 \geq 0$
   $-3x^2 - 4x + 7 \geq 0$

6. $x^2 - x - 6 \leq 0$
   $2x^2 + 7x - 4 \geq 0$

Graph the function and solve these systems of inequalities. Compare the results with the exercises in Paragraph 3-4-2.

7. $7x^2 - 10x - 17 > 0$
   $-x^2 + 2x + 24 > 0$

8. $15x^2 + x - 28 > 0$
   $20x^2 - 31x + 12 > 0$

9. $-12x^2 + 43x - 35 < 0$
   $35x^2 + 24x - 27 < 0$

10. $21x^2 + 26x - 15 < 0$
    $15x^2 - 4x - 35 > 0$

11. $x^2 - 6x - 7 \leq 0$
    $x^2 + x - 6 \leq 0$

12. $5x^2 - 8x + 3 \leq 0$
    $x^2 + 2x - 8 \leq 0$

Exercises on problem solving

13. Same problem as Example 3 with $q_1 = -t^2 + 10t$ and $q_2 = -t^2 + 13t - 22$

14. Same problem as Example 4 with $EL_1 = f(t) = -2t(t-12)$,
    $EL_2 = g(t) = 2(t-12)(t-24)$, $ELO = h(t) = \frac{1}{2}(t-3)(t-21)$

4-4. SOLVING THE QUADRATIC INEQUALITY $y \gtreqless ax^2 + bx + c$

This type of inequality is in the standard form:

$$y \gtreqless f(x) = ax^2 + bx + c \qquad (1)$$

Solving this inequality means finding all points (x , y) in the coordinate plan whose coordinates make the inequality (1) true.
To know which open half plane is the solution set, replace the zero coordinates of the origin into the inequality (1). If it is true then the open half plane containing the origin is the solution set.

Example 1. Solve the inequality $y > f(x) = x^2 - 3x + 2$ (1)

Solution. First, graph the function $y = f(x)$ by setting up the variation table

x	$-\infty$		0		1		$\frac{3}{2}$		2		$+\infty$
y	$+\infty$	↘	2	↘	0	↘	$-\frac{1}{2}$	↗	0	↗	$+\infty$

min.

The graph of this function is shown in Figure 9. Consider a point D (x , y) in the open half plane above the graph. This point has its coordinates (x , y) making the inequality (1) true. The solution set of (1) is the open half plane **above** the graph because any point in this open half plane will have its coordinates making the inequality (1) true. See Figure 9.

Figure 9   Figure 10

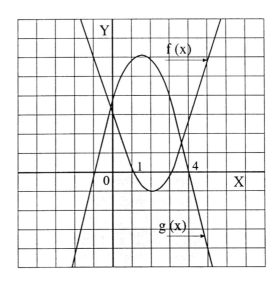

Example 2. Solve the inequality   y < g(x) = -x² + 2x + 3   (2)

Solution. Set up the variation table and graph the function g(x)

x	$-\infty$		-1		0		1		3		$+\infty$
y	$-\infty$	↗	0	↗	3	↗	4	↘	0	↘	$-\infty$

max.

The graph of g(x) is shown in Figure 9.

The solution set is the open half plane which is **below** the graph of g(x).

4-5. SOLVING THE SYSTEM OF QUADRATIC INEQUALITIES
[y ≥ f(x), y ≥ g(x)]

These systems are in the standard form

$$y \geq f(x) = ax^2 + bx + c \quad (1) \quad (a \neq 0)$$
$$y \geq g(x) = dx^2 + ex + f \quad (2) \quad (d \neq 0)$$

To find the solution set of each inequality, use the solving rule by replacing $x = 0$ into the inequalities (1) and (2).

Example 3. Solve the system $y > f(x) = x^2 - 3x + 2$ (1)
$y < g(x) = -x^2 + 2x + 3$ (2)

Solution. First graph the two functions f(x) and g(x)
The solution set of inequality (1) is the open half plane **below** the graph of f(x).
The solution set of the inequality (2) is the open half plane **above** the graph of g(x).
The solution set of the system is the common region shared by both open half planes on Figure 9. The two graphs of f(x) and g(x) are not included in the solution set. They are drawn with dotted lines.

Remark. To find the x-coordinates of the intersections of the two graphs, make f(x) = g(x) and solve the resulting equation for x.

	$x^2 - 3x + 2 = -x^2 + 2x + 3$	Make f(x) = g(x)
or	$2x^2 - 5x - 1 = 0$	Move terms and simplify
Two real roots	$\dfrac{5 \pm 5.74}{4}$	Quadratic formulas
	x = -0.18 and x = 2.68	Answers.

Example 4. Solve the system $y \leq f(x) = x^2 - 4x + 3$ (1)
$y \leq g(x) = -x^2 + 3x + 4$ (2)

Solution. Graph the two functions f(x) and g(x)

The solution set of the inequality (1) is the open half plane **below** the graph of f(x)
The solution set of the inequality (2) is the open half plane **below** the graph of g(x)
The solution set of the system (1) and (2) is the shaded region.
The x-coordinates of the two intersections are the real roots of the equation

$x^2 - 4x + 3 = -x^2 + 3x + 4$ Make f(x) = g(x)

or $2x^2 - 7x - 1 = 0$ Move terms and simplify

Two real roots $\dfrac{7 \pm \sqrt{57}}{4}$ Quadratic formulas

x = 3.64 and x = - 0.14.   Answers.

Note. The two graphs of f(x) and g(x) are both included in the solution set. They are drawn with solid lines. See Figure 10.

Exercises on solving inequalities by using graphs of functions.

Solve these inequalities by using the graph of the function f(x):

1. $y > f(x) = 3x^2 - 5x + 2$ 2. $y < f(x) = 6x^2 - x - 15$

3. $y < f(x) = 3x^2 - 8x - 11$ 4. $y > f(x) = 3x^2 - 6x - 1$

5. $y > -4x^2 + 39x - 56$ 6. $y < 6x^2 - x - 35$

7. $y > 15x^2 + x - 35$ 8. $y < -21x^2 + 13x + 20$

9. $y \geq 7x^2 - 50x + 7$ 10. $y \leq 5x^2 + 9x - 2$

Solve these systems of inequalities:

11. $y < -2x^2 + 3x + 5$ 12. $y > x^2 - 7x + 6$
     $y > x^2 - 4x + 3$ $y < -3x^2 + 5x + 8$

13. $y < -5x^2 + 2x + 7$ 14. $y < -3x^2 + 5x + 8$
     $y < 3x^2 - 4x + 7$ $y > x^2 + x - 6$

15. $\quad y < x^2 + 2x - 8$
    $\quad y < -x^2 + 3x - 10$

16. $\quad y > 6x^2 - x - 15$
    $\quad y > -5x^2 - 9x + 2$

17. $\quad y \leq 3x^2 + 4x - 4$
    $\quad y \geq -2x^2 + 5x + 7$

18. $\quad y \geq 3x^2 - 5x - 8$
    $\quad y \leq -x^2 + 7x - 6$

4-6. REVIEW EXERCISES.

a. Solve these inequalities by graphing:

1. $\quad 2x^2 - 9x + 9 > 0$
2. $\quad -3x^2 + 5x + 12 < 0$
3. $\quad 3x^2 - 4x - 7 < 0$
4. $\quad 2x^2 - 8x + 5 < 0$
5. $\quad 4x^2 - 9x - 9 > 0$
6. $\quad -x^2 + 7x + 8 > 0$
7. $\quad 3x^2 - 4x - 7 \leq 0$
8. $\quad 2x^2 - 8x + 7 \geq 0$

b. Solve by graphing these systems. Compare the results with those in Paragraph 3-7-d

1. $\quad 2x^2 - 8x + 5 > 0$
   $\quad 8x^2 - 2x - 15 < 0$

2. $\quad x^2 - 4x + 2 > 0$
   $\quad 2x^2 - 9x + 9 > 0$

3. $\quad 4x^2 - 9x - 9 < 0$
   $\quad 3x^2 - 20x + 12 < 0$

4. $\quad -3x^2 - x + 14 < 0$
   $\quad x^2 - 7x - 8 < 0$

5. $\quad 5x^2 - 2x - 16 > 0$
   $\quad -7x^2 + 26x - 15 > 0$

6. $\quad 14x^2 - 15x - 9 > 0$
   $\quad x^2 - 7x - 8 < 0$

7. $\quad x^2 - 5x + 5 \geq 0$
   $\quad -7x^2 + 78x - 11 \geq 0$

8. $\quad -3x^2 + 22x - 7 \leq 0$
   $\quad 7x^2 + 9x + 2 \leq 0$

SOLVING QUADRATIC EQUATIONS AND INEQUALITIES
BOOKLET EXERCISE'S ANSWERS

Chapter 1. Solving quadratic equations.

1-3-1. 1. Real roots -2, $\frac{1}{5}$ 3. -4, $\frac{1}{2}$ 5. -1, $-\frac{3}{2}$

     7. $\frac{2}{3}$, $-\frac{3}{7}$ 9. $\frac{-3 \pm \sqrt{2}}{7}$ 11. $\frac{-5 \pm \sqrt{13}}{6}$

1-3-2. 1. Real roots -2, $\frac{2}{3}$ 3. $\frac{5}{2}$ and $-\frac{3}{8}$ 5. -1 and $\frac{11}{3}$

     7. $\frac{-1 \pm \sqrt{6}}{2}$ 9. 3 and $\frac{7}{3}$ 11. $-1 \pm \frac{2\sqrt{10}}{5}$

1-6. 1. $\frac{1}{2}$, -4 3. $-\frac{1}{2}$, $-\frac{5}{3}$ 5. $\frac{3}{4}$, $-\frac{7}{2}$

     7. $\frac{3}{5}$, $-\frac{2}{7}$ 9. -1, $-\frac{3}{2}$ 11. $-\frac{1}{3}$, $\frac{7}{4}$

     13. -2, $-\frac{19}{8}$ 15. $-\frac{3}{8}$, $\frac{5}{2}$ 17. $\frac{2}{3}$, $-\frac{3}{7}$

More exerc. 1. $-\frac{3}{5}$, $-\frac{8}{7}$ 3. $\frac{6}{5}$, $\frac{7}{4}$ 5. $-\frac{5}{3}$, $\frac{8}{5}$

     7. $\frac{2}{3}$, $\frac{3}{7}$ 9. $-\frac{5}{2}$, $\frac{7}{3}$ 11. $\frac{4}{3}$, -2

     13. $-\frac{1}{11}$, 11 15. $\frac{1}{13}$, 17

1-7-1. 1. $63x^2 + 50x - 77 = 0$ 3. $16x^2 - 34x - 15 = 0$

     5. $4x^2 - 39x + 56 = 0$ 7. $7x^2 + 29x + 24 = 0$

     9. $9x^2 - 12x - 1 = 0$ 11. $25x^2 - 20x + 1 = 0$

     13. $16x^2 + 24x + 4 = 0$ 15. $25x^2 - 20x + 13 = 0$

     17. $4x^2 - 12x + 25 = 0$ 19. $16x^2 + 24x + 34 = 0$

1-7-2. Problem 1. Main equation $297x + 297x = 3x^2 - 1200$
$3x^2 - 594x - 1200 = 0$ Answer: 200 km/hr

Problem 3. Main equation $12x^2 - 151x + 475 = 0$ Answers: $\frac{19}{3}$, $\frac{25}{4}$

Problem 5. $12x^2 - 113x + 253 = 0$.   Answers: $\frac{11}{3}$, $\frac{23}{4}$

Problem 7. $21x^2 - 58x + 21 = 0$.   Answers: $\frac{3}{7}$, $\frac{7}{3}$

Problem 9. $9x^2 - 80x - 9 = 0$ Answer: 9

Problem 11. $21x^2 - 148x + 7 = 0$ Answer: $\frac{1}{7}$

Review exercises. 1. $\frac{2}{3}$, -2   3. $\frac{1}{2}$, -4   5. 3 , $\frac{3}{2}$

7. $\frac{2}{3}$, $\frac{11}{5}$ 9. -1 , $\frac{10}{7}$ 11. $\frac{3}{7}$, $-\frac{5}{3}$

13. $\frac{4}{3}$, $-\frac{7}{5}$ 15. $\frac{3}{5}$, $-\frac{9}{7}$ 17. $\frac{1}{5}$, -13

19. $\frac{1}{2}$, $\frac{17}{4}$ 21. $\frac{2 \pm \sqrt{3}}{3}$ 23. $\frac{2 \pm \sqrt{7}}{3}$

Chapter 2. Solving quadratic equations: Special cases

2-1-1. 1. $\frac{3 \pm \sqrt{5}}{2}$ 3. $\frac{5 \pm \sqrt{5}}{2}$

5. $\frac{-3 \pm \sqrt{13}}{2}$ 7. 8 , 7   9. -3 , 13

2-1-3. 1. -3 , -8   3. 2 , 6   5. 7 , 2

7. 2 , -3   9. 8 , -9   11. 2 , -24

13. -2 , 28   15. 4 , 12   17. 2 , -36

2-2. 1. $x = 5$ and $x = -1$ 3. 5 , -3   5. 3 , -7

7.   6 , 2   9. -2 , -8

2-3. 1. $m = \pm 2$ 3. $m = \pm 2\sqrt{2}$ 5. $m = -8$

7. $m = -5$ 9. $p = 1$

2-4. **1.** $x = \pm 1$, $x = \pm\sqrt{6}$ **3.** $x = \pm\sqrt{2}$ **5.** $x = \pm 1$, $x = \pm\dfrac{\sqrt{6}}{3}$

7. $x = \pm \dfrac{2\sqrt{3}}{3}$ **9.** $x = \pm \dfrac{2\sqrt{3}}{3}$ **11.** $x = \pm \dfrac{\sqrt{3}}{2}$, $x = \pm \dfrac{2\sqrt{5}}{5}$

2-5. **1.** Real roots -1, $-\dfrac{3}{2}$ **3.** 1 and $-\dfrac{13}{5}$ **5.** -4 and $\dfrac{1}{2}$

7. -1 and $-\dfrac{1}{8}$ **9.** $\dfrac{1}{2}$ and $-\dfrac{7}{3}$ **11.** $\dfrac{11}{3}$ and $\dfrac{3}{5}$

2-6. **1.** $x = 3$ (Rejected), $x = 7$ (Accepted).   **3.** 18 (R), 3 (A)

5. $\dfrac{4}{3}$ (R), $\dfrac{3}{2}$ (A)   **7.** 5 (R), 7 (A)   **9.** $\dfrac{9}{2}$ (R), 10 (A)

2-8-a. **1.** Real roots 4, 3 **3.** $\dfrac{3 \pm \sqrt{5}}{2}$

5. $\dfrac{3 \pm \sqrt{13}}{2}$ **7.** 9, 2 **9.** 1, 4

2-8-b. **1.** $-1 \pm \sqrt{2}$ **3.** $-4 \pm 2\sqrt{5}$ **5.** $2 \pm 3\sqrt{2}$

7. $-4 \pm 4\sqrt{2}$ **9.** 2, 26

2-8-c. **1.** 1, 5 **3.** $2 \pm \sqrt{13}$

5. 2, 5 **7.** -1, -10

2-8-d. **1.** $m = 1$ **3.** $m = 13$

5. $m = 3$ **7.** $m = 3$, $m = -\dfrac{9}{2}$

2-8-e. **1.** $x = \pm 1$, $x = \pm \dfrac{\sqrt{21}}{3}$ **3.** $\pm \dfrac{\sqrt{39}}{3}$, $\pm \dfrac{\sqrt{2}}{2}$ **5.** ± 2

7. $\pm \dfrac{\sqrt{3}}{3}$, $\pm \dfrac{\sqrt{14}}{2}$ **9.** $\pm\sqrt{3}$, $\pm\sqrt{7}$

2-8-f. **1.** Real roots: $\dfrac{3}{2}$ (Accepted), and 2 (Rejected))   **3.** $-\dfrac{6}{7}$, 2

5. -5, 6 **7.** -10, -11 **9.** $\dfrac{3}{2}$, $\dfrac{1}{7}$

2-8-g. **1.** 3 (R), 7 (A)   **3.** $\dfrac{35}{4}$ (R), 3 (A)   **5.** No real roots

7. -1 (A), 4 (A)   **9.** 6 (A), 5 (A)   **11.** 3 (A), 11 (A)

CHAPTER 3. Solving quadratic inequalities.

3-2. 1. Solution set $(-\infty, -1)$, $(\frac{14}{5}, +\infty)$ 3. $(-11, \frac{1}{3})$

5. $(-\infty, \frac{5}{4})$, $(\frac{3}{2}, +\infty)$ 7. $(-3, \frac{2}{3})$

9. $(-\infty, -\frac{7}{5}]$, $[\frac{4}{3}, +\infty)$ 11. $(-\infty, \frac{3-\sqrt{3}}{2}]$, $[\frac{3+\sqrt{3}}{2}, +\infty)$

3-3. 1. Solution set: $(-\infty, -\frac{7}{2})$, $(-1, +\infty)$ 3. $(\frac{1}{3}, 3)$

5. $(-\infty, \frac{5}{4})$, $(\frac{3}{2}, +\infty)$ 7. $(-\frac{4}{3}, \frac{5}{2})$

9. $(-\infty, -\frac{4}{5}]$, $[-\frac{2}{3}, +\infty)$ 11. $[-3-\frac{\sqrt{2}}{2}, -3+\frac{\sqrt{2}}{2}]$

3-4-1. 1. Solution set: $(1, \frac{8}{3})$ 3. $(-2, -1)$

5. $(-4, -\frac{3}{2})$, $(-1, \frac{1}{2})$ 7. No solution set   9. $[-\frac{7}{5}, \frac{5}{3}]$

3-4-2. 1. $(-4, -1)$, $(\frac{17}{7}, 6)$ 3. $(-\frac{9}{7}, \frac{3}{5})$ 5. $[-1, 2]$

7. $(-\infty, -7)$, $(8, +\infty)$ 9. $(2, 7)$ 11. $[-\frac{5}{3}, -\frac{1}{2}]$

3-5-2. 1. Solution set $(2, \frac{9}{2})$ 3. $(-\frac{3}{5}, \frac{1}{4})$ 5. $(-\frac{9}{7}, -\frac{5}{7})$

3-5-3. 1. Solution set: $(1, 2)$ 3. $(-1, \frac{1}{2})$ 5. $(-\frac{5}{3}, -\frac{1}{2})$

7. $(1, \frac{7}{4})$ 9. Solution set: $(-\infty, -1)$, $(-\frac{1}{3}, 1)$ or $(\frac{7}{4}, +\infty)$

3-7-a. 1. Solution set $(-\frac{7}{3}, \frac{11}{2})$ 3. $(-\infty, \frac{3}{5})$, $(5, +\infty)$ 5. $(\frac{7}{4}, 8)$

7. $(-\infty, -\frac{4}{3})$ or $(3, +\infty)$ 9. $(-\infty, -\frac{3}{8}]$ or $[\frac{5}{2}, +\infty)$

3-7-b. 1. $(-\infty, \frac{5}{4})$, $(2, +\infty)$ 3. $(\frac{6}{7}, 5)$ 5. $(\frac{7}{5}, 6)$

7. $(-\infty, \frac{7}{5})$ or $(2, +\infty)$ 9. $[-\frac{5}{3}, -\frac{1}{2}]$ 11. $[\frac{-2-\sqrt{7}}{3}, \frac{-2+\sqrt{7}}{3}]$

3-7-c. 1. $(\frac{5}{4}, \frac{7}{3})$ 3. $(-\frac{5}{3}, -1)$ 5. $(-\frac{1}{2}, 3)$

7. No solution set   9. $[2+\sqrt{3}, 8]$

3-7-d. 1. $(-\frac{5}{4}, 2 - \frac{\sqrt{6}}{2})$ 3. $(\frac{2}{3}, 3)$

5. $(2, 3)$ 7. $[\frac{1}{7}, \frac{5 - \sqrt{5}}{2}]$, $[\frac{5 + \sqrt{5}}{2}, 11]$

3-7-e. 1. $(-1, \frac{1}{7})$ 3. $(-\infty, -4)$, $(\frac{8}{3}, +\infty)$

3-7-f. 1. $(-1, 1)$, $(2, 5)$ 3. $(-\sqrt{3}, -1)$, $(1, \sqrt{3})$

3-7-g. 1. Write f(x) in the form $f(x) = a[(x + \frac{b}{2a})^2 - \frac{D}{4a^2}]$. When $D < 0$, the term in the bracket is positive. By consequence, f(x) has the same sign as a.

2. When $D = 0$, f(x) is reduced to the form $f(x) = a(x + \frac{b}{2a})^2$. The term in the bracket is positive for all values of x not equal to $-\frac{b}{2a}$. By consequence, f(x) has the same sign as a for all values of x except when $x = -\frac{b}{2a}$.

3. Write f(x) in the form $f(x) = a(x - x_1)(x - x_2)$ and set up the sign table. Suppose $x_1 < x_2$.

x	$-\infty$		x_1		x_2		$+\infty$
$x - x_1$		$-$	0	$+$		$+$	
$x - x_2$		$-$		$-$	0	$+$	
$(x - x_1)(x - x_2)$		$+$	0	$-$	0	$+$	

We see that when x is within the interval (x_1, x_2), f(x) has the opposite sign of a.

Chapter 4. Solving quadratic inequalities by graphing

4-2. See answers on 3-3.

4-3. 1. Solution set $(1, 4)$ 3. $(1, 4)$ 5. $[-\frac{7}{3}, -1]$

7. $(-4, -1), (\frac{17}{7}, 6)$ 9. $(-\frac{9}{7}, \frac{3}{5})$ 11. $[-1, 2]$

13. a. $(2 - 10)$, both stations work
    b. $(0, 2)$ and $(10, 11)$, only one station works
    c. approximately 30 K cubic meter at t = 3.
14. a. (x_1, x_2) water runs from pond to ocean
    b. About 110 cm between 6:00 am to 7:00 am
    c. x_1 = 0:56 am and x_2 = 13:55 pm
    d. During intervals $(0, 0:56$ am$)$ and $(13:55$ pm$, 24)$, water runs from ocean to pond.

4-6-a. 1. $(-\infty, \frac{3}{2})$ or $(3, +\infty)$ 3. $(-1, \frac{7}{3})$

5. $(-\infty, -\frac{3}{4})$, $(3, +\infty)$ 7. $[-1, \frac{7}{3}]$

4-6-b. See 3-7-d.